D0222096

Statistics Principles & Methods, 7e

Student Solutions Manual

Richard A. Johnson – University of Wisconsin at Madison

Gouri K. Bhattacharyya

Prepared by Mark McKibben – West Chester University of Pennsylvania

PUBLISHER	Laurie Rosatone
ACQUISITIONS EDITOR	Joanna Dingle
ASSISTANT EDITOR	Jacqueline Sinacori
SENIOR EDITORIAL ASSISTANT	Courtney Welsh
SENIOR CONTENT MANAGER	Karoline Luciano
SENIOR PRODUCTION EDITOR	Kerry Weinstein

Founded in 1807, John Wiley & Sons, Inc. has been a valued source of knowledge and understanding for more than 200 years, helping people around the world meet their needs and fulfill their aspirations. Our company is built on a foundation of principles that include responsibility to the communities we serve and where we live and work. In 2008, we launched a Corporate Citizenship Initiative, a global effort to address the environmental, social, economic, and ethical challenges we face in our business. Among the issues we are addressing are carbon impact, paper specifications and procurement, ethical conduct within our business and among our vendors, and community and charitable support. For more information, please visit our website: www.wiley.com/go/citizenship.

Copyright © 2014 John Wiley & Sons, Inc. All rights reserved. No part of this publication may be reproduced, stored in a retrieval system, or transmitted in any form or by any means, electronic, mechanical, photocopying, recording, scanning or otherwise, except as permitted under Sections 107 or 108 of the 1976 United States Copyright Act, without either the prior written permission of the Publisher, or authorization through payment of the appropriate per- copy fee to the Copyright Clearance Center, Inc., 222 Rosewood Drive, Danvers, MA 01923 (Web site: www.copyright.com). Requests to the Publisher for permission should be addressed to the Permissions Department, John Wiley & Sons, Inc., 111 River Street, Hoboken, NJ 07030-5774, (201) 748-6011, fax (201) 748-6008, or online at: www.wiley.com/go/permissions.

ISBN 978-1-118-61631-4

Printed in the United States of America

10 9 8 7 6 5 4 3 2 1

Table Contents

Chapter 1

INTRODUCTION

1.1 The *statistical population* consists of the answers provided by all men to whether they wait until Valentine's day or the day before to purchase flowers. The *sample* is the responses from 451 men.

1.3 The *statistical population* consists of the answers of all college students when asked their number of close friends. The *sample* is the answers given by the twenty college students.

1.5 The *statistical population* is the responses to the questions are you more stressed than last year. The *sample* is the responses of the 40 adults.

1.7 No, those who have a higher rate of drug use are arguably more likely to visit the first aid station, and so the percentage may not be representative of the rate of drug use for all attendees of the rave.

1.9 (a) anecdotal – This number was simply tabulated from members of a class. The students were not randomly selected and the information as to whether or not they sent a text message was recorded.

**NOTE: If the actual class was randomly selected as part of a larger observational study, then it *would be* sample-based.

(b) anecdotal – This is the opinion of one person.

(c) sample-based – A small number of college-age students were selected to answer the given question.

1.11 Answers will vary.

1.13 The notion of "comfortable" is variable depending on the user (e.g., hand size and button positioning will affect one's impression of how comfortable it is to use a mouse).

Better statement: Determine if consumers prefer a new style mouse to the mouse they currently use.

1.15 Purpose: Determine the amount of time it takes those who use the Internet to make hotel reservations in San Francisco.

1.17 At the lab, receptionist and x-ray.

1.19 The *population* consists of the entire set of responses from all teenagers, 13 to 17 years old, in the United States while the *sample* consists of the responses of the particular 1055 teens contacted in the telephone survey.

1.21 (a) A person living in Chicago is the *unit*.

(b) The *variable of interest* is the characterization of a Chicago resident as eligible to vote or not.

(c) The *statistical population* consists of the collection of voter eligibility characterizations of each resident of the city of Chicago.

1.23 (a) An individual golfer is the *unit*.
(b) The *statistical population* consists of the collection of all possible decisions regarding "choice of hole" made by all golfers. The *sample* consists of the set of responses for this particular group of 46 golfers.

1.25 No, because a self-selection bias is likely to exist since only people who are interested in this particular exam are likely to answer, and such people perceive such a problem with the values.

1.27 The newspaper is suggesting that the *population* is the collection of preferences for each adult in the city while the *sample* is the collection of preferences of the particular persons who sent in their votes. This sample is apt to be non-representative because those persons in the sample are self-selected. Only the few who feel very strongly positive will likely send in a vote.

1.29 (a) This is anecdotal. No data given.
(b) The yes/no answer regarding multiple credit cards, for each of the 22 students, is the sample on which the statement is based.
(c) The yes/no answer regarding destination outside the continental United States, for each of the 55 people at the airport, is the sample on which the statement is based.

1.31 The term "too long" is not well defined. By asking a number of people, we may determine that 5 minutes is too long. Further, the time will not be the same for all people. One improved statement of the purpose is:

Purpose: Determine if over half the persons take over 5 minutes to get cash during the lunch hour.

1.33 First number the classrooms 1 to 35. In Table 1, we started in row 20 using columns 29 and 30. Reading downward, and ignoring 00 and numbers above 35, we selected rooms 8, 7, 1, and 9. Answers will vary.

1.35 We started in row 10 and read down column 9 and then down column 6 from the top. We ignored the second digit in a pair, and kept reading, where the two digits were the same. That type of assignment of students is not allowed. Answers will vary.

20 pairs of random digits

4,0	1,2	4,2	5,2	2,1
5,1	3,2	2,0	7,6	5,4
0,1	2,5	2,7	3,6	5,4
2,4	5,7	3,5	6,7	7,2

(a) $4/20 = 0.20$
(b) $9/20 = 0.45$
(c) $7/20 = 0.35$

1.37 (a) The miniature poodles could never be observed even if they greatly outnumber the Great Danes. Only the big dogs can volunteer to show they were inside the fence.

(b) Persons who call-in their opinions are self-selected because they have strong opinions. This is analogous to the big dogs who are the only volunteers to show they were inside the fence.

Chapter 2

ORGANIZATION AND DESCRIPTION OF DATA

2.1 (a) The percentage in other classes is $100 - 28.5 - 13.9 - 13.4 - 12.4 - 9.0 = 22.8$ %

(b)

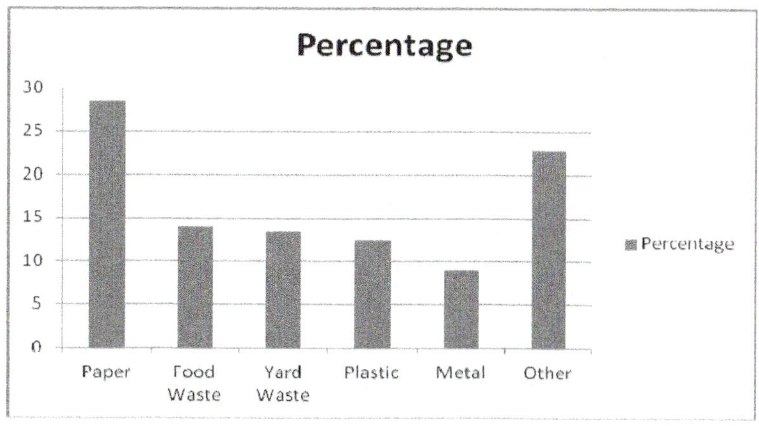

A frequency table for each type of waste is as follows:

Type of Waste	Frequency (in millions of tons)
Paper and Paperbound	72.39
Food	35.31
Yard	34.04
Plastic	31.50
Metal	22.86
Other	57.91

(c) The percentage of waste that is paper and paperboard is: 28.5 %

The percentage of waste in the top two categories is: $28.5 + 22.8 = 51.3\%$

The percentage in the top five categories is:

$$28.5+22.8+13.9+13.4+12.4=91\%$$

2.3 The frequency table for number of activities is

Number of Activities	Frequency	Relative Frequency
0	7	7/40 = 0.175
1	10	10/40 = 0.25
2	13	13/40 = 0.325
3	5	5/40 = 0.125
4	2	2/40 = 0.05
5	1	1/40 = 0.025
6	1	1/40 = 0.025
7	1	1/40 = 0.025
Total	40	1.00

This is the relative frequency histogram:

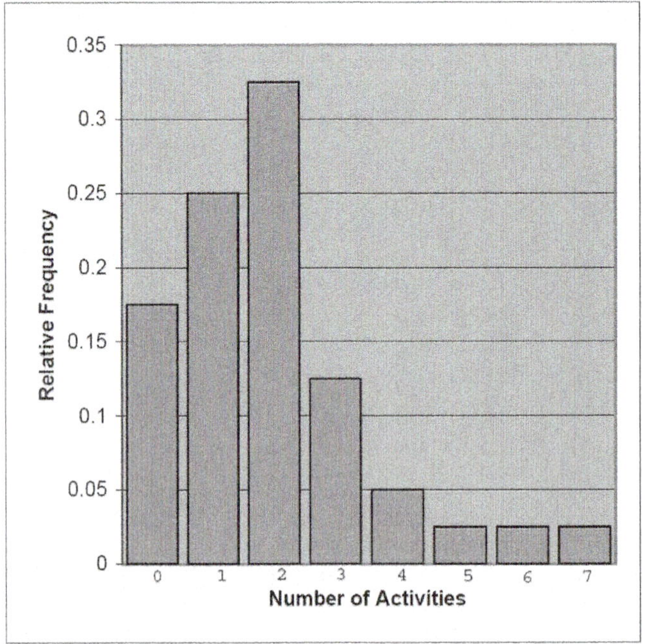

2.5 (a) The table of relative frequencies for workers in the department is

Mode of Transportation	Frequency	Relative Frequency
Drive alone	25	$25/40=0.625$
Car pool	3	$3/40=0.075$
Ride bus	7	$7/40=0.175$
Other	5	$5/40=0.125$
Total	40	1.000

(b) The pie chart for workers in the department is

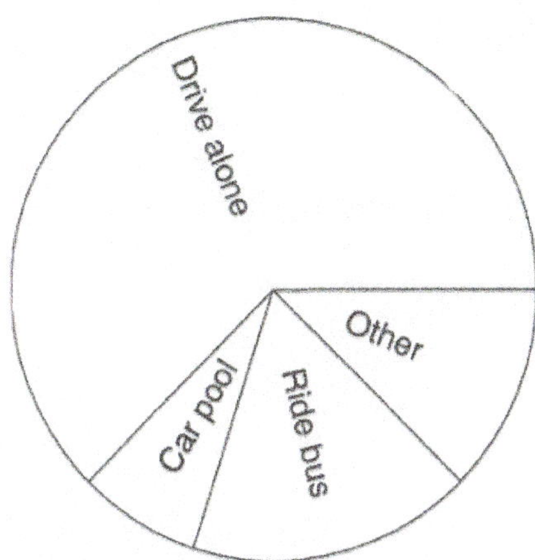

2.7 There are overlapping classes in the grouping. A report of 3 stolen bicycles will fall in two classes.

2.9 There is a gap. The response 5 close friends does not fall in any class. The last class should be 5 or more.

2.11 (a) Yes. (b) Yes. (c) Yes. (d) No. (e) No.

2.13 (a) The relative frequencies are 0.18, 0.48, 0.26, and 0.08 for 0, 1, 2, and 3 bags, respectively.

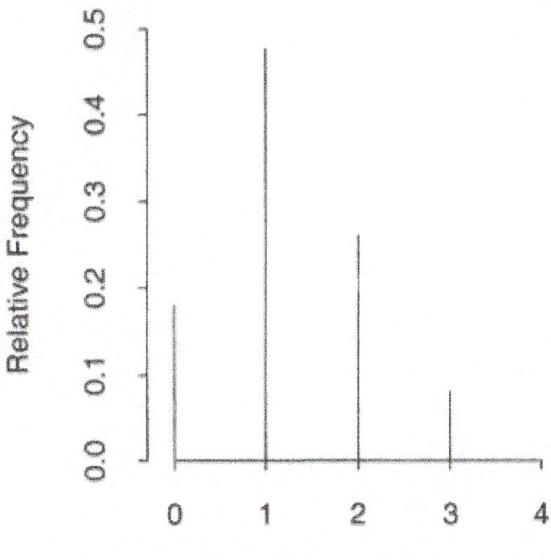

(b) Nearly one-half of the passengers check exactly one bag. The longest tail is to the right.

(c) The proportion of passengers who fail to check a bag is $9/50 = 0.18$.

2.15 The dot diagram of amounts of radiation leakage is

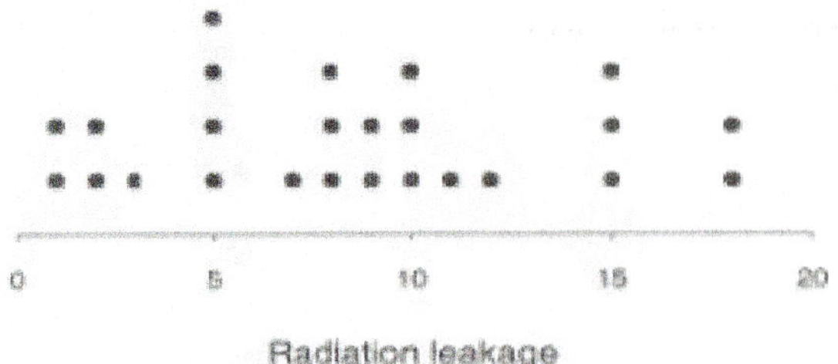

2.17 (a) The dot diagram of number of CFUs is

(b) There is a long tail to the right with one extremely large value of 1600 CFU units.

(c) There is one day so the proportion is $1/15 = 0.067$

2.19 (a) In the following frequency distribution of lizard speed (in meters per second), the left endpoint is included in the class interval but not the right endpoint.

Class Interval	Frequency	Relative Frequency
0.45 to 0.90	2	0.067
0.90 to 1.35	6	0.200
1.35 to 1.80	11	0.367
1.80 to 2.25	5	0.167
2.25 to 2.70	6	0.200
Total	30	1.001 (rounding error)

(b) All of the class intervals are of length 0.45 so we can graph rectangles whose heights are the relative frequency. The histogram is

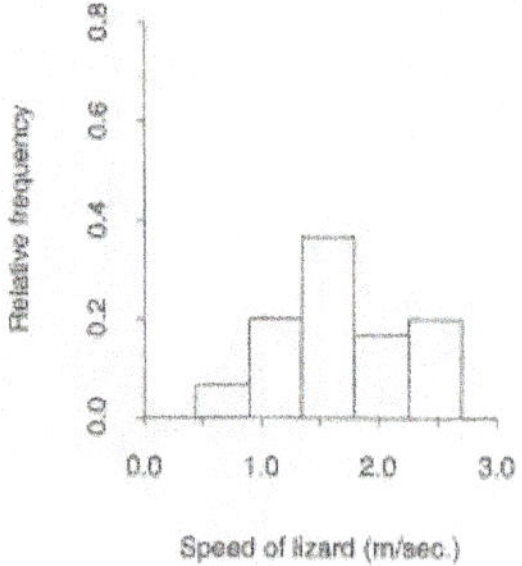

2.21 This time, the frequency distribution is given by

Class Interval	Frequency
(6.0, 6.3]	12
(6.3, 6.6]	16
(6.6, 6.9]	10
(6.9, 7.2]	11
(7.2, 7.9]	8
Total	57

The corresponding frequency histogram is as follows:

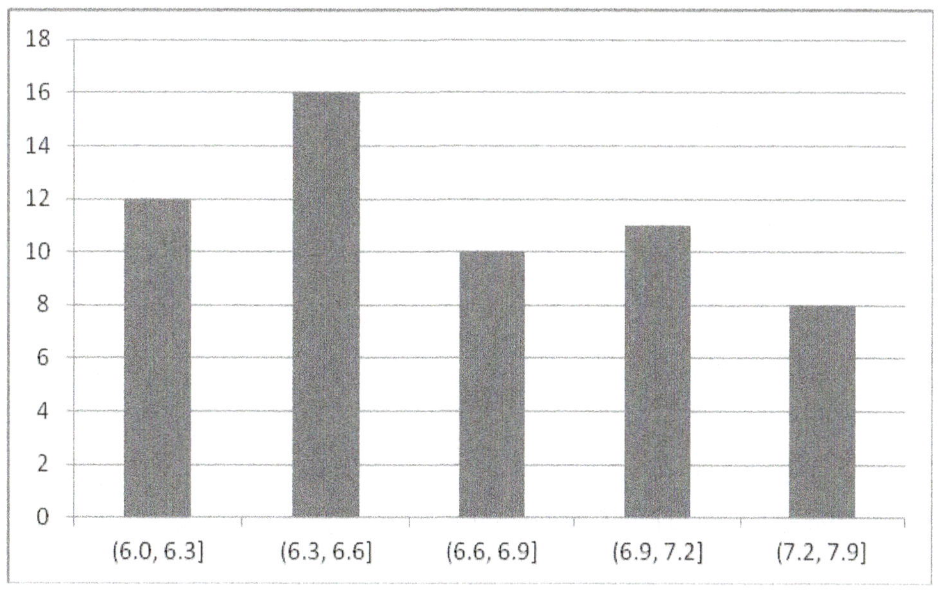

2.23 The stem-and-leaf display of the amount of iron present in the oil is

0	6
1	2234455567777889
2	000000222445567799
3	022444566
4	1167
5	12

2.25 The double-stem display of the amount of iron present in the oil is

0	6
1	22344
1	55567777889
2	00000022244
2	5567799
3	022444
3	566
4	11
4	67
5	12

2.27 The five-stem display of the Consumer Price Index in 2001 for the given cities is

20	0289
21	01124799
22	055
23	12333
24	448
25	3

2.29 (a) The median is 3. The sample mean is

$$\bar{x} = \frac{2+5+1+4+3}{5} = \frac{15}{5} = 3$$

(b) The mean is

$$\bar{x} = \frac{26+30+38+32+26+31}{6} = \frac{183}{6} = 30.5$$

The ordered measurements are: 26, 26, 30, 31, 32, 38

$$\text{median} = \frac{30+31}{2} = 30.5$$

(c) The sample mean is

$$\bar{x} = \frac{-3+0-2-1+2-3}{7} = -1$$

The ordered measurements are: $-3, -3, -2, -1, 0, 0, 2$.

The median is -1.

2.31 (a) $\bar{x} = 3810/15 = 254$.

(b) The ordered observations are:

10 20 50 60 80 90 90 110
140 180 260 340 380 400 1600

So, the median is 110 CFU units. The one very large observation makes the sample mean much larger. Hence, the sample median is better to use in this instance.

2.33 The mean is $956/12 = 79.67$. The claim ignores variability and is not true. It is certainly unpleasant with a daily maximum temperature $105^{o}F$ in July.

2.35 (a) The sample mean is computed by adding all twenty-five data points and then dividing that sum by 25. Doing so yields 8.48.

(b) The data is fairly spread out with five very small values and five comparatively large numbers. In such case, since there are no extremely small or large outliers that would unreasonably throw off the mean, the mean would be an appropriate measure here.

2.37 The mean, 118.05, is one measure of center tendency and the median, 117.00, is another. The value 118.05 tells us that, on average, a baby weighed 118.05 ounces. The median tells us that about half of the babies weighed at least 117 ounces while roughly half weighed at most 117 ounces.

2.39 (a) $\bar{x} = \dfrac{1(7)+2(9)+3(6)+4(5)+5(3)}{30} = 2.6$ (returns)

(b) Sample median is 2 returns.

2.41 (a) $\bar{x} = 271/40 = 6.775$ days.

(b) Sample median $= (6+7)/2 = 6.5$. Both the sample mean and the sample median give a good indication of the amount of mineral lost.

2.43 Sample median $= (176+187)/2 = 181.5$ (minutes).

2.45 (a) The dot diagram for the diameters (in feet) of the Indian mounds in southern Wisconsin is

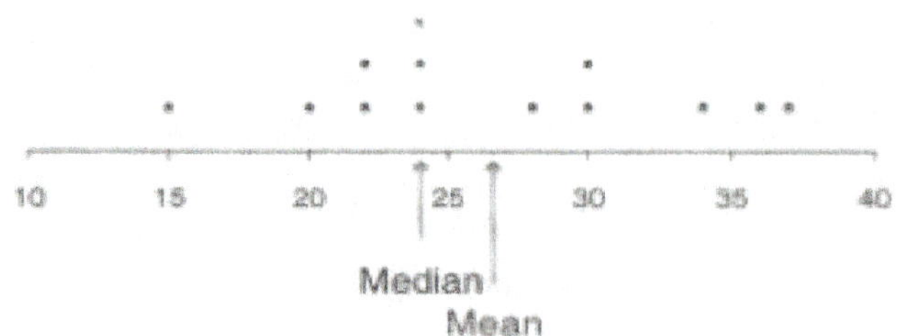

(b) $\bar{x} = 346/13 = 26.62$. Sample median $= 24$.
(c) $13/4 = 3.25$, so we count in 4 observations. $Q_1 = 22$ and $Q_3 = 30$.

2.47 (a) Median $= (152+154)/2 = 153$.
(b) $40/4 = 10$, so we need to count in 10 observations. The 11-th smallest observation also satisfies the definition. This yields $Q_1 = \dfrac{135+136}{2} = 135.5$.

Using a similar approach, we find that $Q_3 = \dfrac{166+167}{2} = 166.5$.

2.49 The ordered data are

50	57	68	69	72	73	73	80	82	91
92	93	94	96	96	100	102	104	105	106
108	109	118	118	127					

Since the number of observations is 25, the median or second quartile is the 13th ordered observation in the list. The first quartile is the 7th ordered observation and the third quartile is the 19th ordered observation:

$$Q_1 = 73 \quad Q_2 = 94 \quad Q_3 = 105$$

2.51 (a) The ordered observations are

10	20	50	60	80	90	90	110
140	180	260	340	380	400	1600	

Since the sample size is 15, the median is the 8th ordered observation 110. To obtain Q_1, we find $15/4 = 3.75$ so the first quartile is the 4th ordered observation in the ordered list. To obtain Q_3, we find 0.75(15)=11.25, so that the third quartile is the 12th ordered observation:

$$Q_1 = 60 \quad Q_3 = 340$$

(b) The 90th percentile requires us to count in at least $0.9(15) = 13.5$ or 14 observations. The 90th sample percentile $= 400$.

2.53 (a) The ordered data are 73, 74, 76, 76, 80. The median is $76^\circ F$ and the mean is $\bar{x} = 379/5 = 75.8^\circ F$.

(b) The mean of $({}^{\circ}F - 32)$ is $\overline{x} - 32$ by property (i) of Exercise 2.52 with $c = -32$. By property (ii)

$$\text{mean of } \frac{5}{9}({}^{\circ}F - 32) = \frac{5}{9}(\text{mean of } ({}^{\circ}F - 32))$$

$$= \frac{5}{9}(\overline{x} - 32) = \frac{5}{9}(75.8 - 32) = 24.33^{\circ}C$$

By similar properties for the median

$$\text{median of } \frac{5}{9}({}^{\circ}F - 32) = \frac{5}{9}(\text{median of } ({}^{\circ}F) - 32) = \frac{5}{9}(76 - 32) = 24.44^{\circ}C$$

2.55 (a)

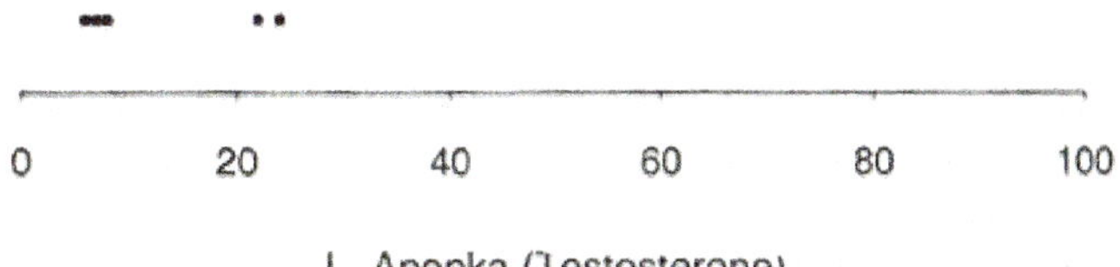

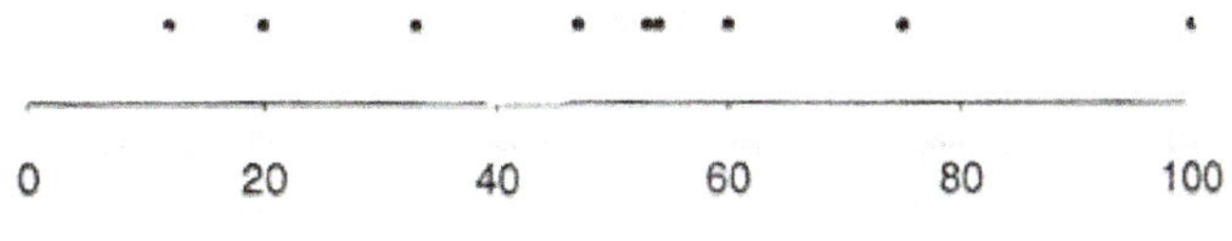

(b)

Lake Apopka $\overline{x} = \dfrac{67}{5} = 13.40$ Lake Woodruff $\overline{x} = \dfrac{454}{9} = 50.44$

(c) From the dot diagrams, the males in Lake Apopka have lower levels of testosterone and their sample mean is only about one-third of that for males in (un-contaminated) Lake Woodruff. This finding is consistent with the environmentalists' concern that the contamination has affected the testosterone levels and the reproductive abilities.

2.57 (a) We carry out all necessary calculations in the following table. The mean is $\bar{x} = 15/3 = 5$.

	x	$x-\bar{x}$	$(x-\bar{x})^2$
	8	3	9
	3	−2	4
	4	−1	1
Total	15	0.0	14

(b) The variance and the standard deviation are

$$s^2 = \frac{14}{3-1} = 7 \quad , \quad s = \sqrt{7} \approx 2.646$$

2.59 (a) We carry out all necessary calculations in the following table. The mean is $\bar{x} = 32/4 = 8$.

	x	$x-\bar{x}$	$(x-\bar{x})^2$
	8	0	0
	6	−2	4
	14	6	36
	4	−4	16
Total	32	0.0	56

(b) The variance and the standard deviation are

$$s^2 = \frac{56}{4-1} = 18.667 \quad , \quad s = \sqrt{18.667} = 4.320$$

2.61 We carry out all necessary calculations in the following table.

	x	x^2
	8	64
	3	9
	4	16
Total	15	89

The variance is

$$s^2 = \frac{1}{n-1}\left(\sum x^2 - \frac{\left(\sum x\right)^2}{n}\right) = \frac{1}{2}\left(89 - \frac{15^2}{3}\right) = \frac{1}{2}(89-75) = 7$$

2.63 (a) $s^2 = (34 - 12^2/5)/4 = 1.30$.

(b) $s^2 = (19 - (-7)^2/6)/5 = 2.167$.

(c) $s^2 = (499 - 59^2/7)/6 = 0.286$.

2.65 $s = \sqrt{(9726 - 346^2/13)/12} = 6.5643$.

2.67 (a) $s^2 = (3{,}140{,}900 - 3810^2/15)/14 = 155{,}225.7$.
(b) $s = \sqrt{155225.7} = 393.99$.
(c) $s^2 = (580900 - 2210^2/14)/13 = 17{,}848.9$, So, $s = \sqrt{17848.9} = 133.6$. The single very large value greatly inflates the standard deviation.

2.69 (a) $\bar{x} = 1862/10 = 186.2$.
(b) $s^2 = (353796 - 1862^2/10)/9 = 787.96$.
(c) $s = \sqrt{787.96} = 28.07$.

2.71 (a) Median $= 68.4$.
(b) $\bar{x} = 478.4/7 = 68.343$.
(c) $s^2 = (32730.34 - 478.4^2/7)/6 = 5.853$. Hence $s = 2.419$.

2.73 (a) The measure of variation displayed is 15.47, the sample standard deviation. The sample variance is $s^2 = 15.47^2 = 239.321$.

(b) The interquartile range is $Q_3 - Q_1 = 131.00 - 106.00 = 25.00$. This means the center half of the data span an interval of length 25 ounces.

(c) Any value smaller than 15.47 would correspond to smaller variation.

2.75 Using the data set in Exercise 2.22, in Exercise 2.47, we determined that $Q_1 = 135.5$ and $Q_3 = 166.5$. Hence,

$$\text{Interquartile range} = Q_3 - Q_1 = 166.5 - 135.5 = 31.0 \text{ points.}$$

2.77 No. Typically, the middle half of a data set is much more concentrated than the combination of the two quarters, one in each tail. As an example, for the water quality data of Exercise 2.17, the range is $1600 - 10 = 1590$ because of one extremely large observation. From the quartiles determined in Exercise 2.51, the interquartile range is $340 - 60 = 280$. The range is six times larger than the interquartile range.

2.79 (a) $\bar{x} = 6.775$ and $s = \sqrt{19.4096} = 4.406$.
(b) The proportion of the observations are given in the following table:

	$\bar{x} \pm s$	$\bar{x} \pm 2s$	$\bar{x} \pm 3s$
Interval:	(2.369, 11.181)	(−2.037, 15.587)	(−6.443, 19.993)
Proportion:	$26/40 = 0.65$	$38/40 = 0.95$	$40/40 = 1.00$
Guidelines:	0.68	0.95	0.997

(c) We observe a good agreement with the proportions suggested by the empirical guideline.

2.81 (a) $\bar{x} = 2.6$ and $s = \sqrt{1.69655} = 1.3025$.

(b) The proportion of the observations are given in the following table:

	$\bar{x} \pm s$	$\bar{x} \pm 2s$	$\bar{x} \pm 3s$
Interval:	(1.2975, 3.9025)	(−0.005, 5.205)	(−1.3075, 6.5075)
Proportion:	$15/30 = 0.50$	$30/30 = 1.00$	$30/30 = 1.00$
Guidelines:	0.68	0.95	0.997

(c) We observe a good agreement with the proportions suggested by the empirical guideline.

2.83 (a) $z = \dfrac{102 - 118.05}{15.47} = -1.037$ (b) $z = \dfrac{144 - 118.05}{15.47} = 1.677$

2.85 For males, the minimum and the maximum horizontal velocity of a thrown ball are 25.2 and 59.9 respectively. The quartiles are:

$$Q_1 = (38.6 + 39.1)/2 = 38.85,$$
$$\text{median} = (45.8 + 48.3)/2 = 47.05,$$
$$Q_3 = (49.9 + 51.7)/2 = 50.8.$$

For females, the minimum and the maximum horizontal velocity of a thrown ball are 19.4 and 53.7 respectively. The quartiles are

$$Q_1 = 25.7, \text{ median} = 30.3, Q_3 = 33.5.$$

The boxplot of the male and female throwing speed are

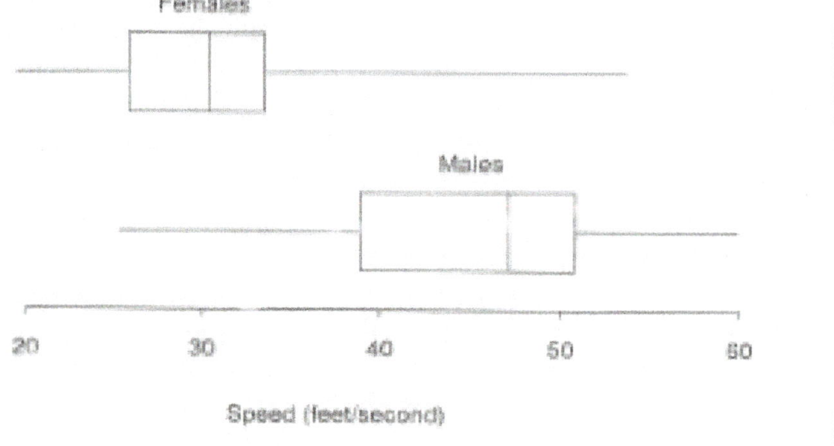

Comparing the two boxplots, we can see that males throw the ball faster than females.

2.87 (a)

$$\bar{x} \approx 21.917 \quad \text{and} \quad s = 5.40$$

(b) Since $\bar{x} - 2s = 21.917 - 2(5.40) = 11.117$ and $\bar{x} + 2s = 21.917 + 2(5.40) = 32.717$ only one of the differences lies outside the interval. The proportion $23/24 = 0.958$ lies within the interval.

2.89 From Exercise 2.38, we know that

$$\bar{x} = 1.925 \quad \text{and} \quad s = \sqrt{\frac{249 - (77)^2/40}{40-1}} = 1.607$$

2.91 (a) The ordered data are

-8
-5
4
5
8
12
15
26
30
47
48
48
52
63

Median $= (15 + 26)/2 = 20.5$ seats lost.

(b) The maximum number of seats lost, 63, occurred when Obama was President. The minimum number, -8 or a gain, occurred during G.W. Bush's term as President.

(c) range $= 63 - (-8) = 71$

2.93

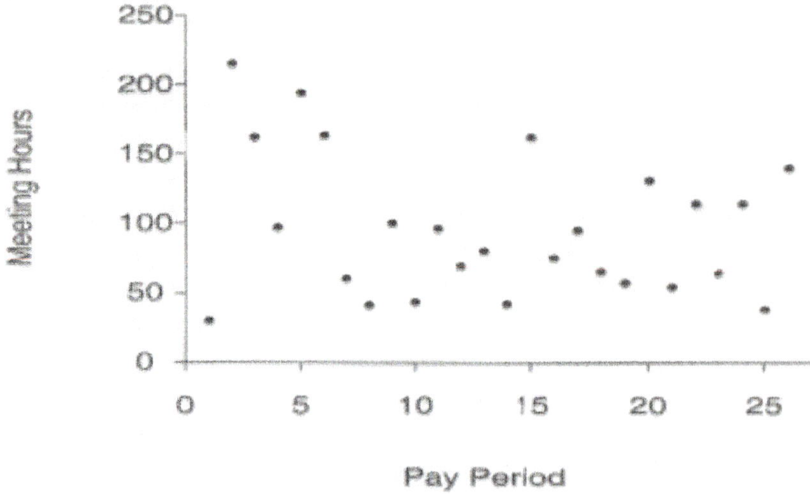

The value 215 from the second pay period looks high and 194 from the fifth period is possibly high.

2.95 We calculate $\bar{x} = 2501/26 = 96.2$ and $s = \sqrt{65254/25} = 51.1$ so the upper limit is $\bar{x} + 2s = 198.4$ and the lower limit is $\bar{x} - 2s = -6.0$ which we take as 0.

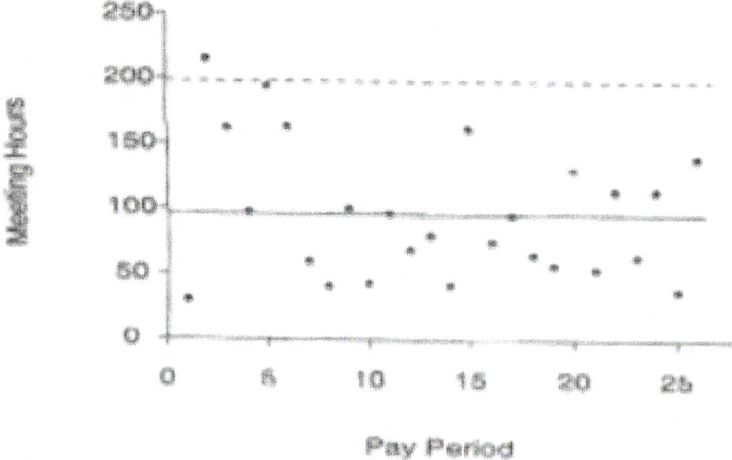

Only the value 215 from the second pay period is out of control.

2.97 (Note: The years have been coded using integers, starting with 1995 as 1 and proceeding consecutively.)

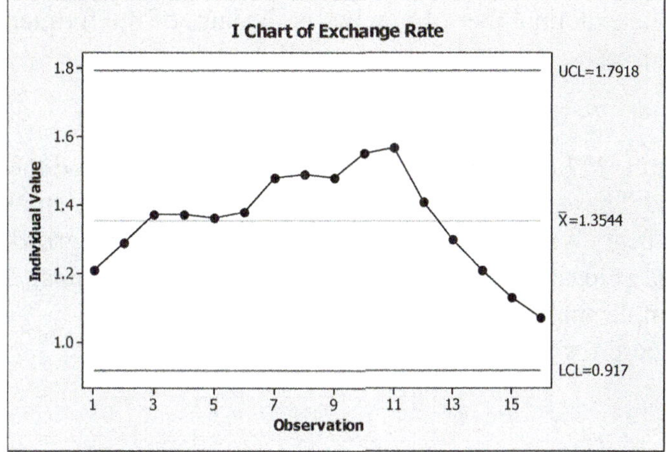

The process appears to be in statistical control.

2.99 (a) The relative frequencies of the occupation groups are:

	Relative Frequency	
	2010	2000
Goods Producing	0.117	0.161
Service (Private)	0.736	0.702
Government	0.147	0.136
Total	1.000	0.999

(b) The proportions of persons in private service occupations and government has increased while the proportion in goods producing have decreased from 2000 to 2010.

2.101 The dot diagrams of heights for the male and female students are

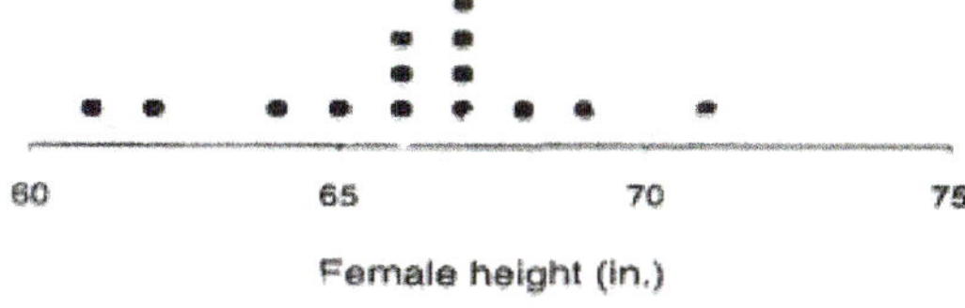

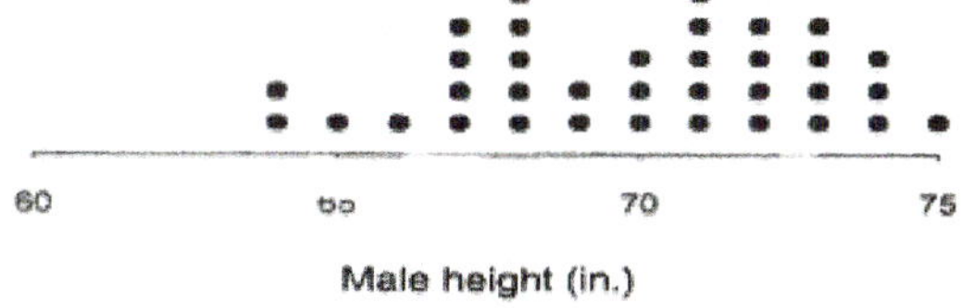

2.103 (a) Yes. The exact number of lunches is the sum of the frequencies of the first three classes.
(b) Yes. The exact number of lunches is the sum of the frequencies of the last three classes.
(c) No.

2.105 (a) The mean, 227.4, is one measure of center tendency and the median, 232.5, is another. These values may be interpreted as follows. On average, the 20 grizzly bears weigh 227.4 pounds apiece. Half of the grizzly bears sampled weighed at least 232.5 pounds while half weighed at most 232.5 pounds.
(b) The sample standard deviation is 82.7 pounds.
(c) The z score for a weight of 320 pounds is

$$z = \frac{320 - 227.4}{82.7} = 1.12$$

2.107 (a) Sample median $= (9+9)/2 = 9$.
(b) $\bar{x} = 271/30 = 9.033$.
(c) The sample variance is

$$s^2 = \frac{1}{29}\left(2561 - \frac{271^2}{30}\right) = 3.895.$$

2.109 (a) $\bar{x} = 7$, $s = 2$
(b) By the properties, the new data set $x + 100$ has sample mean $= (7+100)$ $= 107$ and standard deviation 2. By direct calculation, we verify

$$\bar{x} = \frac{106+108+104+109+108}{5} = 107$$

$$s^2 = \frac{(106-107)^2 + (108-107)^2 + (104-107)^2 + (109-107)^2 + (108-107)^2}{4} = 4$$

(c) By the properties, the new data set $-3x$ has sample mean $= -3(7) = -21$ and standard deviation $|-3|s = 3(2) = 6$. By direct calculation, we verify

$$\bar{x} = \frac{-18-24-12-27-24}{5} = -21$$

$$s^2 = \frac{(3)^2 + (-3)^2 + (9)^2 + (-6)^2 + (-3)^2}{4} = 9\left(\frac{1+1+9+4+1}{4}\right) = 36$$

2.111 (a) The dot diagrams are

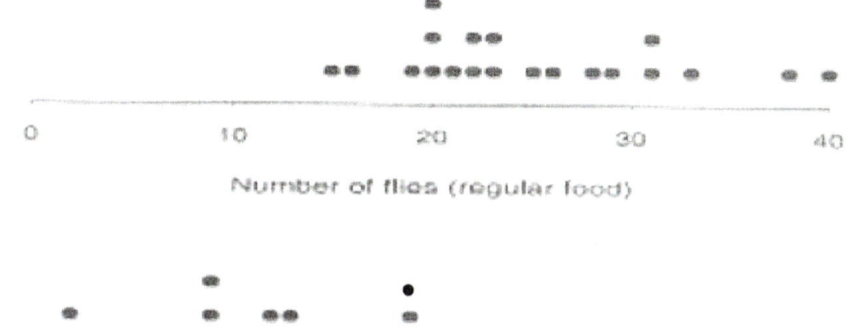

(b) From the dot diagrams we can see the number of flies (grape juice) is centered at about 11 and the number of flies (regular food) is centered near 25. The spread looks about the same.

(c) Regular food: $\bar{x} = 25.1$, $s = 6.84$.
Grape juice: $\bar{x} = 11.05$, $s = 6.194$.

2.113 (a) $\bar{x} = 5.38$ and $s = 3.42$.
(b) Median $= 5$.
(c) Range $= 13 - 0 = 13$.

2.115 (a) Median $= 4.505$, $Q_1 = 4.30$ and $Q_3 = 4.70$.
(b) 90th percentile $= = (4.80 + 5.07)/2 = 4.935$.
(c) $\bar{x} = 4.5074$ and $s = 0.368$.
(d) The boxplot of acid rain in Wisconsin is

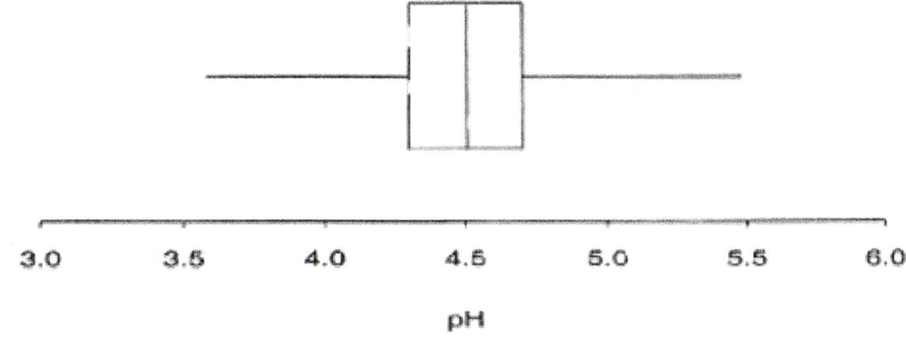

2.117 (a) Median $= 6.7$, $Q_1 = 6.4$ and $Q_3 = 7.1$.
(b) $\bar{x} = 6.7386$ and $s = 0.4632$.
(c) The boxplot of the data is

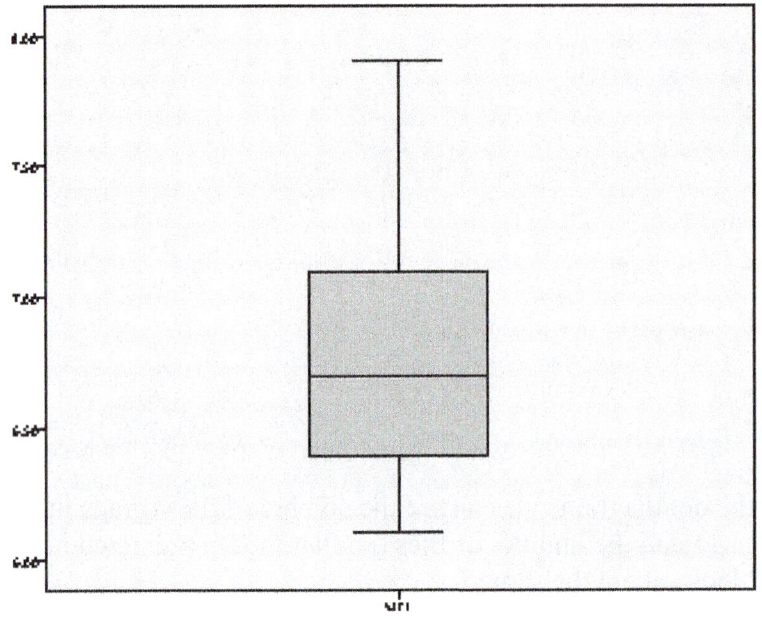

2.119 (a)

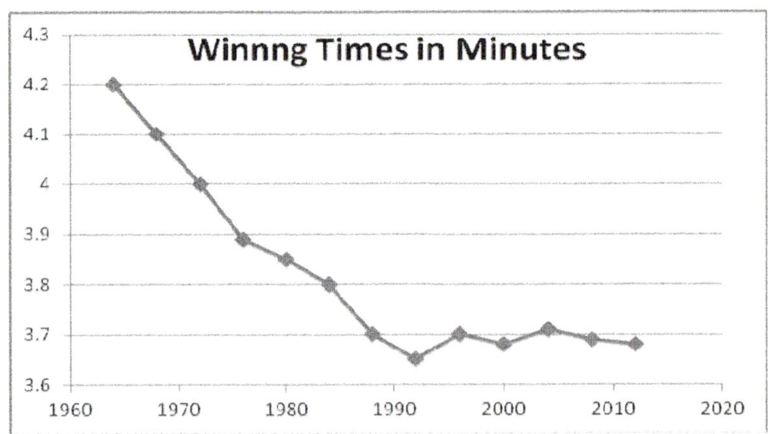

(b) It is not reasonable, because a frequency distribution would not show the systematic decrease of the winning times over the years, which is the main feature of these observations.

2.121 Using Excel, we obtain

$$\bar{x} \approx 80.30 \quad \text{and} \quad s \approx 39.30 .$$

2.123 (a) The partial MINITAB output is

Variable	N	Mean	Median	StDev
Speed	30	1.724	1.665	0.573

Variable	Minimum	Maximum	Q1	Q3
Speed	0.500	2.670	1.288	2.125

(b) The partial MINITAB output for the acid rain data.

	N	MEAN	MEDIAN	STDEV
PH	50	4.5074	4.5050	0.3681

	MIN	MAX	Q1	Q3
PH	3.5800	5.4800	4.2950	4.7000

(Note that MINITAB uses a slightly different convention for determining Q_1 and Q_3.)

2.125 The mean is 7.181 and the standard deviation is 1.282.

2.127 The mean and standard deviation given by MINITAB are the rounded off values of the answer given by SAS.

2.129 (a) The histogram of the alligator data is

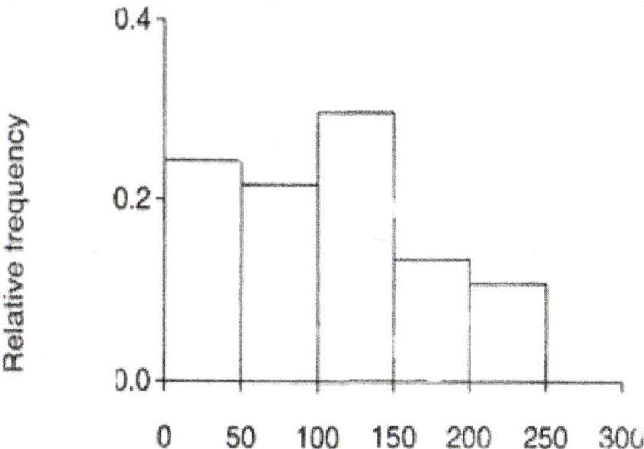

(b)

$$\bar{x} = \frac{4035}{37} = 109.1 \quad s = \sqrt{\frac{155672}{37-1}} = 65.8$$

2.131 (a)

```
Descriptive Statistics: Maltextr

Variable          N       Mean      Median     StDev
Maltextr         40     77.458      77.400     1.101
```

(b) The ordered observations are

75.3	75.7	75.9	75.9	76.2	76.3	76.4	76.4	76.6	76.6
76.7	76.9	76.9	77.0	77.0	77.1	77.4	77.4	77.4	77.4
77.4	77.5	77.6	77.6	77.8	77.9	77.9	77.9	77.9	77.9
78.0	78.1	78.3	78.4	78.4	78.5	79.1	79.2	80.0	80.4

There are 40 observations so the median $= (77.4+77.4)/2 = 77.4$. The first quartile is the average of the $40/4 = 10$th and 11th observations in the sorted list. $Q_1 = (76.6+76.7)/2 = 76.65$ and $Q_1 = (77.9+78.0)/2 = 77.95$.

(c) The interval $\bar{x} \pm s$ or $(76.36, 78.56)$ has relative frequency $30/40 = 0.75$ compared to 0.683. The interval $\bar{x} \pm 2s$ or $(75.26, 79.66)$ has relative frequency $38/40 = 0.95$ compared to 0.95. The interval $\bar{x} \pm 3s$ or $(74.16, 80.76)$ has relative frequency 1 compared to 0.997. The agreement is quite good.

Chapter 3

DESCRIPTIVE STUDY OF BIVARIATE DATA

3.1 (a) The table, with completed marginal totals, is:

	Degree of Nausea				
	None	Slight	Moderate	Severe	Total
Pill	43	36	18	3	100
Placebo	19	33	36	12	100
Total	62	69	54	15	200

(b) The relative frequencies, by row, are

	Degree of Nausea				
	None	Slight	Moderate	Severe	Total
Pill	0.43	0.36	0.18	0.03	1.00
Placebo	0.19	0.33	0.36	0.12	1.00

(c) A much higher proportion, 0.43, of pill takers avoided nausea as compared to the proportion 0.19 among those who took the placebo. Also the proportion of persons suffering moderate and severe nausea was much lower among those receiving the pill.

3.3 The relative frequencies, by row, are

	10 or less	More than 10	Total
Biology	0.40	0.60	1.000
Physical	0.30	0.70	1.000
Social	0.52	0.48	1.000

A larger percentage of physical science and biology majors study longer for their final exams that do social science majors.

3.5 (a) The two-way frequency table is:

		Iron		
		Low	High	Total
Alkalinity	Low	8	2	10
	High	4	5	9
Total		12	7	19

(b) The relative frequencies are:

		Iron		
		Low	High	Total
Alkalinity	Low	0.421	0.105	0.526
	High	0.211	0.263	0.474
Total		0.632	0.368	1.000

(c) The relative frequencies, by row, are:

		Iron		
		Low	High	Total
Alkalinity	Low	0.800	0.200	1.000
	High	0.444	0.556	1.000

3.7 (a) The two-way frequency table is:

	Major				
	B	H	P	S	Total
Male	12	4	5	14	35
Female	6	0	4	4	14
Total	18	4	9	18	49

(b) The relative frequencies are:

	Major				
	B	H	P	S	Total
Male	0.245	0.082	0.102	0.286	0.715
Female	0.122	0	0.082	0.082	0.286
Total	0.367	0.082	0.184	0.368	1.001

Alternate solution using Minitab:

When a data set is large, it is useful to enter it once on a computer and instruct it to do the counting. To do so, we encode gender and intended major as numbers. We choose 0 if male, 1 if female and

$$1 = B\ ,\ 2 = H\ ,\ 3 = P\ ,\ 4 = S$$

With the coded data in a file called 2.94.dat.

```
ROWS: GENDER      COLUMNS: MAJOR

          1       2       3       4     ALL

0        12       4       5      14      35
1         6       0       4       4      14
ALL      18       4       9      18      49

CELL CONTENTS --
                COUNT
```

We can also calculate relative frequencies by row. More precisely, 100 × (relative frequency) is obtained from the MINITAB command

```
TABLE C1 C4;
ROWPERCENT.
```

```
ROWS: GENDER      COLUMNS: MAJOR

          1        2        3        4       ALL
0     34.29    11.43    14.29    40.00    100.00
1     42.86       --    28.57    28.57    100.00
ALL   36.73     8.16    18.37    36.73    100.00

CELL CONTENTS --
                 % OF ROW
```

3.9 (a) The proportions are calculated by row $0.548 = 23/42$ and so on.

	Male	Female	Total
English	0.548	0.452	1.000
Computer science	0.844	0.156	1.000

(b) There appears to be gender bias, favoring males, in the Computer Science department. The relative frequencies in the English department do not indicate the obvious presence of bias.

3.11 (a) The proportions, by row, for each condition are

Good Condition

	Died	Survived	Total
Research hospital	0.021	0.979	1.000
Community hospital	0.027	0.973	1.000

Bad Condition

	Died	Survived	Total
Research hospital	0.050	0.950	1.000
Community hospital	0.070	0.930	1.000

(b) The research hospital has a higher proportion of patients in good condition that survive, 0.979 vs.0.973, and a higher proportion of patients in poor condition that survive, 0.950 vs. 0.930. Whether you are in bad or in good condition, you should prefer the research hospital.

(c) We have reached just the opposite conclusion of that reached in Exercise 3.10. In this example of Simpson's paradox, the condition of the patient acted as the lurking variable. The proportion of patients in poor condition is much higher at the research hospital so that kept down their overall survival rate calculated in Exercise 3.10.

3.13 (a) Of course, the fact that 21 out of 57, or proportion .368 quit smoking, by itself, would seem to be stronger evidence. Intuitively, we tend to think incorrectly that no persons would have quit without the medicated patch.

(b) Most people respond positively when they are given attention. The placebo trials make it possible to treat all subjects alike except for the presence or absence of medication. Twenty percent, 11 out of 55, responded positively to the procedure, even without the medication. This makes the success of the medicated patch less spectacular but provides a proper frame of reference.

3.15 (a) Positive – more sales persons should be able to see more people and sell more real estate.

(b) Positive – in general better players get paid higher salaries.

(c) Positive – one would expect sales to increase with the amount of TV advertising of the cola.

(d) Negative – strength diminishes with age after middle age.

3.17 No. The value of r can be small even is there is a strong relationship along a curve as illustrated in Figure 2 of the text.

3.19 (a) A computer calculation gives $r = -0.460$ for males.
The scatter plot diagram for males and the multiple scatter plot are

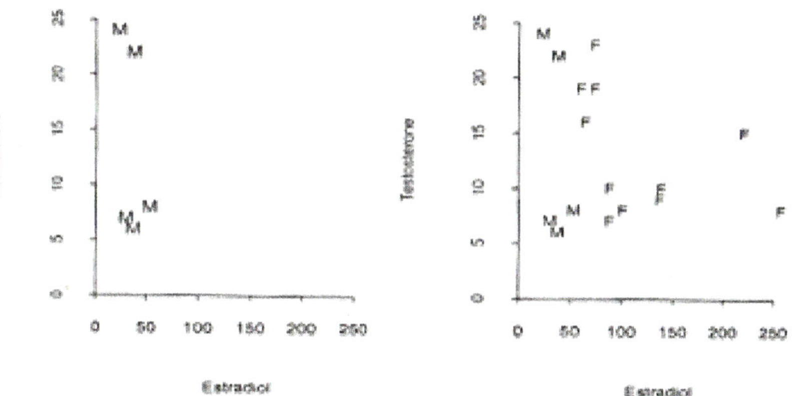

(b) A computer calculation gives $r = -0.415$ for females.

(c) Both have about the same testosterone levels but Females have higher levels of estradiol and are more variable.

3.21 Only Figure 8(c) has a northwest-southeast pattern indicating a negative value for r. Since the tightest pattern, indicating the largest r, is in Figure 8(a) the matches are
(a) $r = -0.3$ and Figure 8(c) (b) $r = 0.1$ and Figure 8(b)
(c) $r = 0.9$ and Figure 8(a)

3.23 Identifying the sums of squares about the means as S_{xx}, S_{yy}, and S_{xy} respectively, we find

$$r = \frac{S_{xy}}{\sqrt{S_{xx}}\sqrt{S_{yy}}} = \frac{-204.3}{\sqrt{530.7}\sqrt{235.4}} = -0.578$$

3.25 Let $x =$ temperature and $y =$ maximum Ozone. Then, with $n = 23$, we calculate

$$\sum x = 1{,}883, \quad \sum y = 1{,}237$$

$$\sum x^2 = 156{,}663 \quad \sum y^2 = 71{,}671 \quad \sum xy = 104{,}221$$

so

$$S_{xx} = \sum x^2 - \frac{(\sum x)^2}{n} = 156{,}663 - \frac{3{,}545{,}689}{24} = 8925.958$$

$$S_{yy} = \sum y^2 - \frac{(\sum y)^2}{n} = 71{,}671 - \frac{1{,}530{,}169}{24} = 7913.958$$

$$S_{xy} = \sum xy - \frac{(\sum x)(\sum y)}{n} = 104{,}221 - \frac{1883 \times 1237}{24} = 7168.042$$

Consequently,

$$r = \frac{S_{xy}}{\sqrt{S_{xx}}\sqrt{S_{yy}}} = \frac{7168.042}{\sqrt{8925.958}\sqrt{7913.958}} = 0.853$$

3.27 (a) The scatter diagram is shown below. The pattern runs from lower left to upper right and is not very tight. We estimate $r = 0.2$.

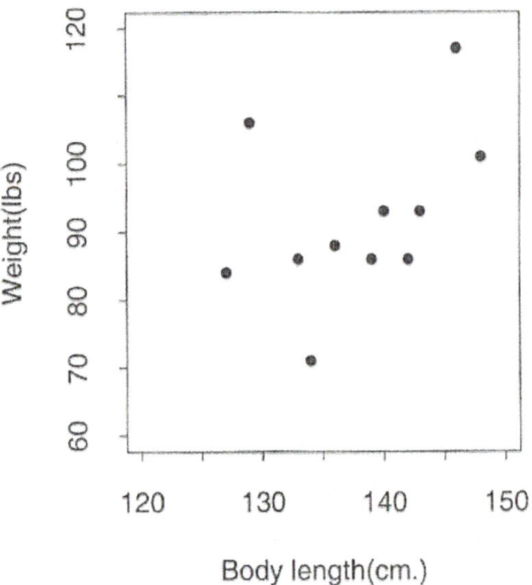

(b) Let $x =$ length and $y =$ weight. We calculate

$$\sum x = 1511 \quad , \quad \sum y = 1011$$
$$\sum x^2 = 208{,}153 \quad \sum y^2 = 94{,}453 \quad \sum xy = 139{,}141$$

so

$$S_{xx} = \sum x^2 - \frac{(\sum x)^2}{n} = 208{,}153 - \frac{(1511)^2}{11} = 596.545$$
$$S_{yy} = \sum y^2 - \frac{(\sum y)^2}{n} = 94{,}453 - \frac{(1011)^2}{11} = 1532.909$$
$$S_{xy} = \sum xy - \frac{(\sum x)(\sum y)}{n} = 139{,}141 - \frac{(1511)(1011)}{11} = 266.364$$

Consequently,

$$r = \frac{S_{xy}}{\sqrt{S_{xx}}\sqrt{S_{yy}}} = \frac{266.364}{\sqrt{596.545}\sqrt{1532.909}} = 0.279$$

(c) The multiple scatter diagram reveals different patterns and one possible F outlier.

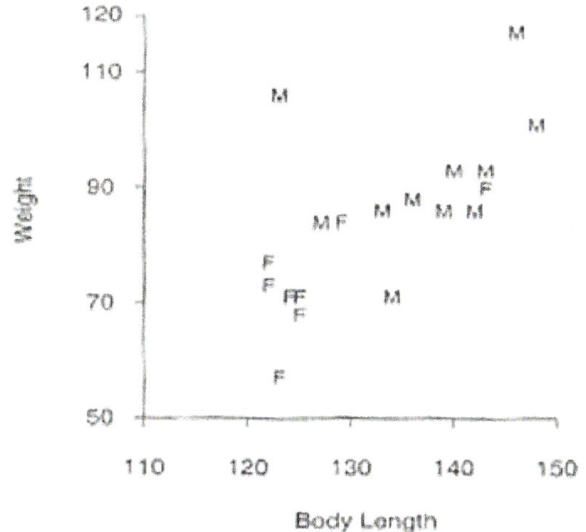

3.29 (a) The scatter diagram is

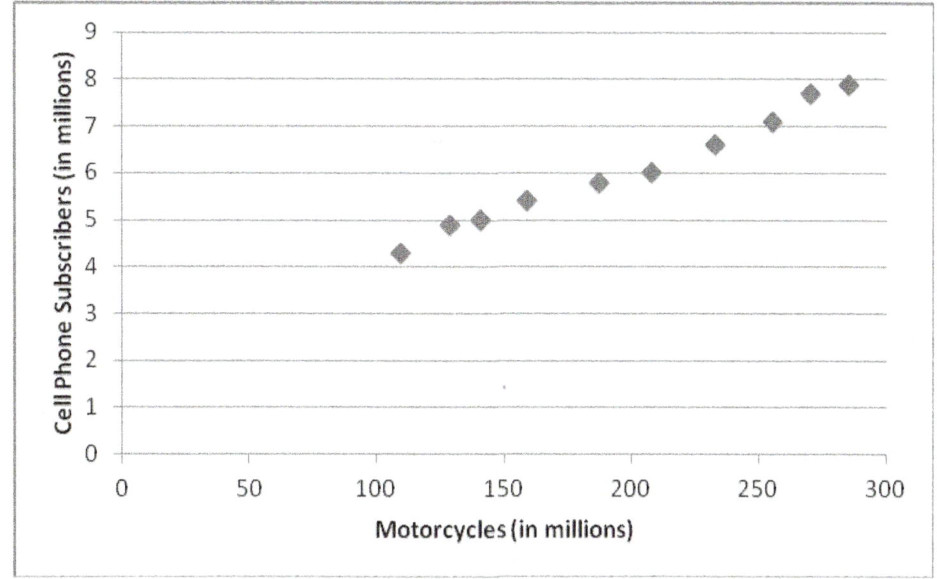

(b) The scatter diagram exhibits a strong correlation but it is hard to imagine any causal relationship between an increase in motorcycle registration and an increase in cell phone usage, so we suspect the presence of lurking variables. A steadily increasing population seems a likely culprit.

3.31 (a) The scatter diagram is shown below.

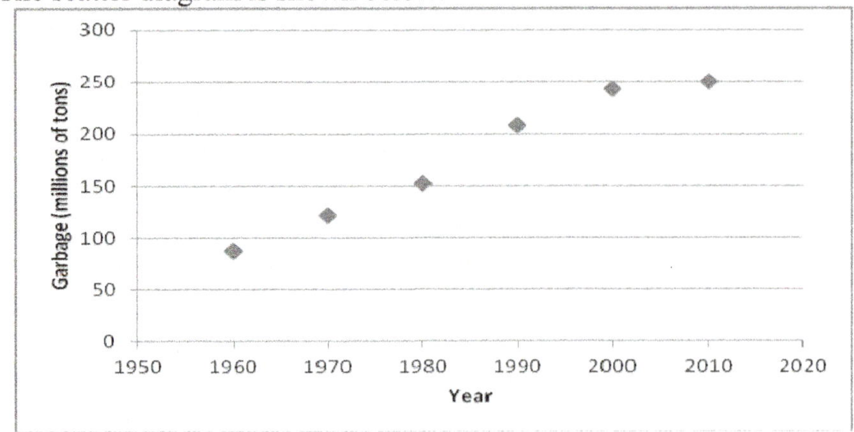

(b) There is a tight southwest to northeast pattern indicating a strong positive correlation. The later years have the largest garbage values.

(c) Population size is also increasing and most likely, even with more recycling, more people mean more garbage.

3.33 (a) Let $x = (\text{year} - 1960)$ and $y =$ amount of garbage (mil.tons). Then, with $n = 6$, we calculate

$$\sum x = 150 \quad , \quad \sum y = 1,062$$

$$\sum x^2 = 5,500 \quad \sum y^2 = 210,302 \quad \sum xy = 32,710$$

so

$$S_{xx} = \sum x^2 - \frac{(\sum x)^2}{n} = 1,750$$

$$S_{yy} = \sum y^2 - \frac{(\sum y)^2}{n} = 22,328$$

$$S_{xy} = \sum xy - \frac{(\sum x)(\sum y)}{n} = 6,160$$

Consequently,

$$r = \frac{S_{xy}}{\sqrt{S_{xx}}\sqrt{S_{yy}}} = \frac{6,160}{\sqrt{1,750}\sqrt{22,328}} = 0.9855$$

(b) The correlation is still 0.9855. Since year is a linear transformation of (year – 1960)

$$\text{year} = 1 \cdot (\text{year} - 1960) + 1960$$

The deviation for each year, or $\left(\text{year} - \overline{\text{year}}\right)$ equals the same deviation for $\left(\text{year} - \overline{1960}\right)$ as you may verify. Consequently the sum of squares for years and the sum of cross-products remain the same and the correlation is unchanged (see Exercise 3.30).

3.35 The value of y at $x=1$ is $2+3(1)=5$ and the value at $x=4$ is $2+3(4)=14$. The line is shown below. The intercept is 2, the value of y at $x=0$, and the slope is 3, the coefficient of x.

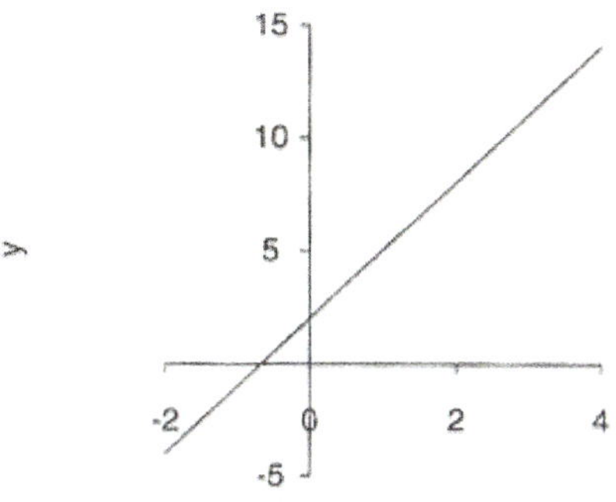

3.37 (a) $y=10(41)-155=255$.

(b) Note that $y=0$ if $10x=155$ or if $x=15.5$. A profit will be made if 16 or more units are sold.

3.39 (a) (b) The scatter diagram and the visually drawn dotted line are shown below.

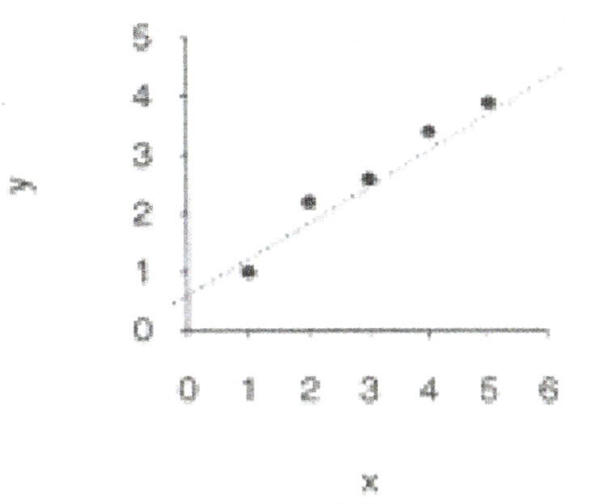

(a) and (b) Visual

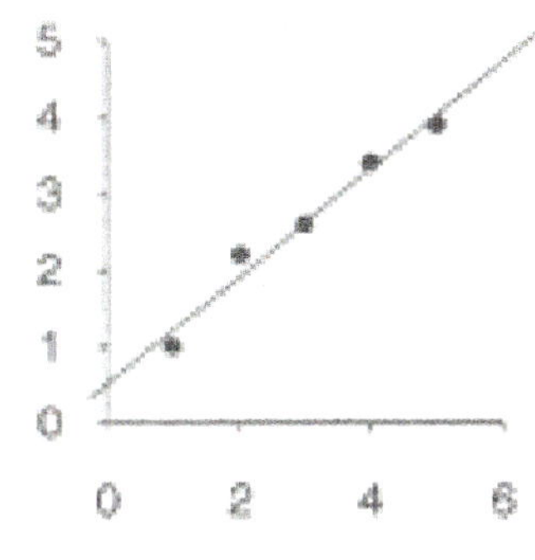

(c) Least Squares

(c) We use the alternative form of calculation

x	y	xy	x^2
1	1.0	1.0	1
2	2.2	4.4	4
3	2.6	7.8	9
4	3.4	13.6	16
5	3.9	19.5	25
15	13.1	46.3	55
$\bar{x}=3$	$\bar{y}=2.62$	$\sum xy$	$\sum x^2$

so

$$S_{xy}=\sum xy-\frac{(\sum x)(\sum y)}{n}=46.3-\frac{(15)(13.1)}{5}=7.00$$

$$S_{xx}=\sum x^2-\frac{(\sum x)^2}{n}=55-\frac{(15)^2}{5}=10.00$$

and

$$\hat{\beta}_1=\frac{S_{xy}}{S_{xx}}=\frac{7}{10}=0.70$$

$$\hat{\beta}_0=\bar{y}-\hat{\beta}_1\bar{x}=2.62-(0.70)3=0.52$$

and the least squares line is $\hat{y}=0.52+0.70x$. This is the solid line in the figure.

3.41 (a) $\hat{\beta}_1=\frac{S_{xy}}{S_{xx}}=\frac{10.2}{9.4}=1.085,\qquad \hat{\beta}_0=\bar{y}-\hat{\beta}_1\bar{x}=\frac{39.9}{9}-(1.085)\frac{19}{9}=2.143$

(b) The least squares fitted line is $\hat{y}=2.143+1.085x$. So, $\hat{y}=2.143+1.085(3)=5.398$ or about 5.4 minutes.

3.43 (a) The scatter diagram is shown below.

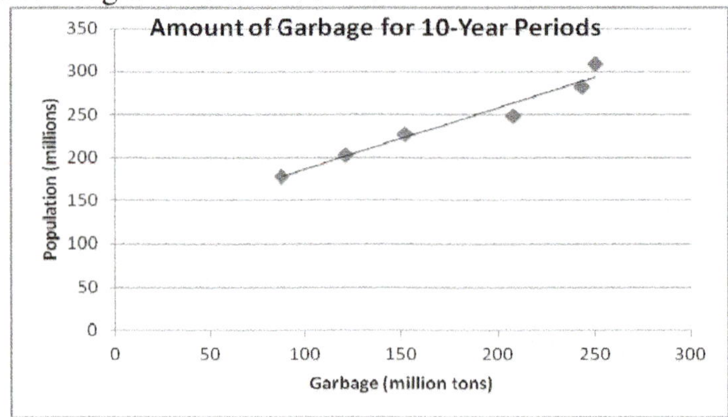

(b) We calculate

x	y	$x-\bar{x}$	$y-\bar{y}$	$(x-\bar{x})(y-\bar{y})$	$(x-\bar{x})^2$
88	179	-89	-62.5	5562.5	7921
121	203	-56	-38.5	2156	3136
152	227	-25	-14.5	362.5	625
208	249	31	7.5	232.5	961
243	282	66	40.5	2673	4356
250	309	73	67.5	4927.5	5329
1062	1449	0	0	15,914	22,328

$$n = 6,\ \sum x = 1062,\ \sum y = 1449,\ \bar{x} = 177,\ \bar{y} = 241.5$$
$$S_{xx} = 22{,}328,\ S_{xy} = 15{,}914$$

Using this, we have

$$\hat{\beta}_1 = \frac{S_{xy}}{S_{xx}} = \frac{15{,}914}{22{,}328} = 0.713 \quad \hat{\beta}_0 = \bar{y} - \hat{\beta}_1\bar{x} = 241.5 - (0.713)(177) = 115.3$$

and so, the least squares line is $\hat{y} = 115.3 + 0.713x$.

(c) The slope of the least squares line says that 0.713 million tons of garbage is created for each 1 million people. Hence, each person generates about 0.7 ton of garbage per year.

3.45 (a) $\bar{x} = 98/30 = 3.3$

(b) The relative frequencies, by manufacturer, are:

	Carbohydrates		
Manufacturer	Above mean	Below mean	Total
General Mills	2	8	10
Kellogg	5	5	10
Quaker	4	6	10
Total	19	11	30

(c)

	Carbohydrates			**Relative Frequency**
Manufacturer	Above mean	Below mean	Total	
General Mills	2	8	10	**1/3**
Kellogg	5	5	10	**1/3**
Quaker	4	6	10	**1/3**
Total	19	11	30	
Relative Frequency	**19/30**	**11/30**	**1.00**	

The row proportions are exactly the same for each row.

3.47 (a) The frequency table is

Size	Drive 2-Wheel	4-Wheel	Total
Small	12	23	35
Full	20	25	45
Total	32	48	80

(b) The relative frequencies are:

Size	Drive 2-Wheel	4-Wheel	Total
Small	0.15	0.2875	0.4375
Full	0.25	0.3125	0.5625
Total	0.40	0.60	1.00

(c)

Size	Drive 2-Wheel	4-Wheel	Total
Small	0.343	0.657	1.000
Full	0.444	0.556	1.000

(d) A larger proportion of small truck purchasers prefer 4-wheel drive.

3.49 (a) Negative; typically, the more time spent on the computer the fewer hours available for friends and other activities.
(b) Somewhat negative; most students cram for finals and the more exams the more late night studying during finals and the fewer hours of sleep.
(c) No relation.
(d) Positive; higher temperature tends to make people more thirsty.

3.51 No. The severity of the forest fire would necessitate more fire fighters to get the fire under control. But, with increased severity comes increased loss.

3.53 (a) The scatter diagram is shown below.

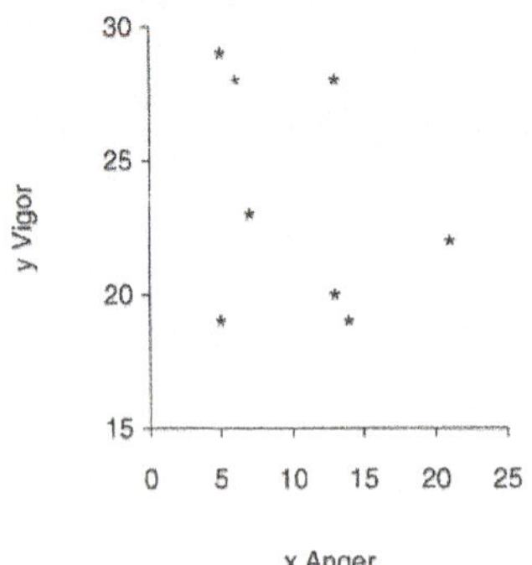

(b) With $n = 8$, we calculate

$$\sum x = 84 \quad , \quad \sum y = 188$$
$$\sum x^2 = 1110 \quad \sum y^2 = 4544 \quad \sum xy = 1921$$

so

$$S_{xx} = \sum x^2 - \frac{\left(\sum x\right)^2}{n} = 1110 - \frac{(84)^2}{8} = 228.00$$
$$S_{yy} = \sum y^2 - \frac{\left(\sum y\right)^2}{n} = 4544 - \frac{(188)^2}{8} = 126.00$$
$$S_{xy} = \sum xy - \frac{\left(\sum x\right)\left(\sum y\right)}{n} = 1921 - \frac{(84)(188)}{8} = -53.00$$

Consequently,

$$r = \frac{S_{xy}}{\sqrt{S_{xx}}\sqrt{S_{yy}}} = \frac{-53}{\sqrt{228.0}\sqrt{126.0}} = -0.313.$$

(c)

$$\hat{\beta}_1 = \frac{S_{xy}}{S_{xx}} = \frac{-53}{228.0} = -0.232$$
$$\hat{\beta}_0 = \bar{y} - \hat{\beta}_1 \bar{x} = \frac{188}{8} - (-0.232)\frac{84}{8} = 25.9$$

and the least squares line is $\hat{y} = 25.9 - 0.232x$.

(d) The equation is $\hat{y} = 25.9 - 0.232(8) = 24.0$. In later chapters, we will see that this least squares fitted line does not have much predictive power because of the small value of r.

3.55 (a) (b) The scatter plot and visually drawn line are shown below.

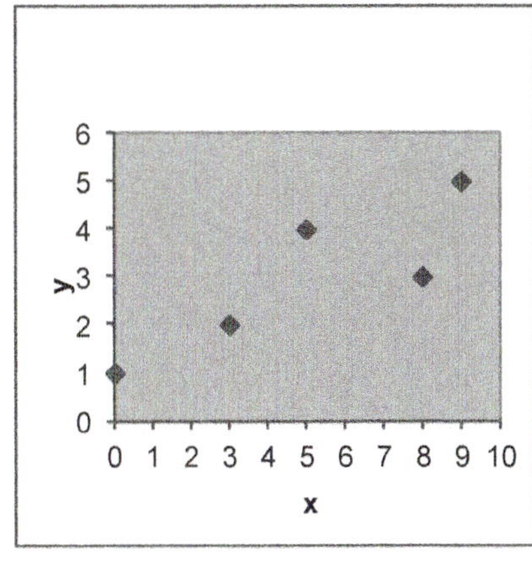

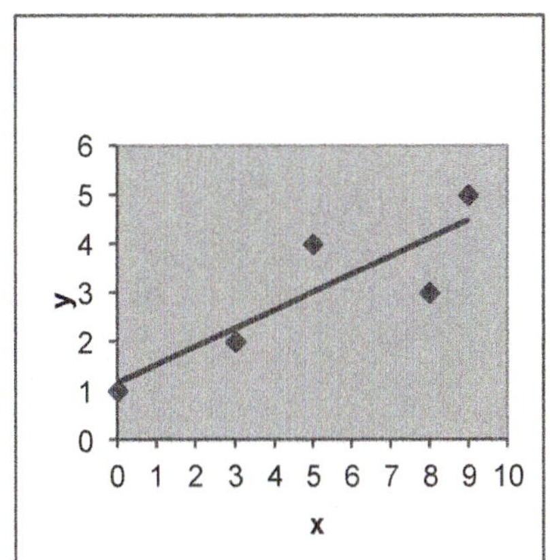

(c) We calculate

x'	y'	$x'-\overline{x'}$	$y'-\overline{y'}$	$\left(x'-\overline{x'}\right)\left(y'-\overline{y'}\right)$	$\left(x'-\overline{x'}\right)^2$
0	1	−5	−2	10	25
3	2	−2	−1	2	4
5	4	0	1	0	0
8	3	3	0	0	9
9	5	4	2	8	16
25	15	0	0	20	54
$\overline{x'}=5$	$\overline{y'}=3$			$S_{x'y'}$	$S_{x'x'}$

so

$$\hat{\beta}_1 = \frac{S_{xy}}{S_{xx}} = \frac{20}{54} = 0.370$$

$$\hat{\beta}_0 = \overline{y} - \hat{\beta}_1 \overline{x} = 3 - (0.370)5 = 1.15$$

and the least squares line is $\hat{y} = 1.15 + 0.370x$. This is the solid line in the figure.

3.57 (a) $x =$ road roughness and $y =$ gas consumption.
(b) $x =$ number of wins and $y =$ total sales.
(c) $x =$ trip distance and $y =$ number of weekends at home.

3.59 (a)

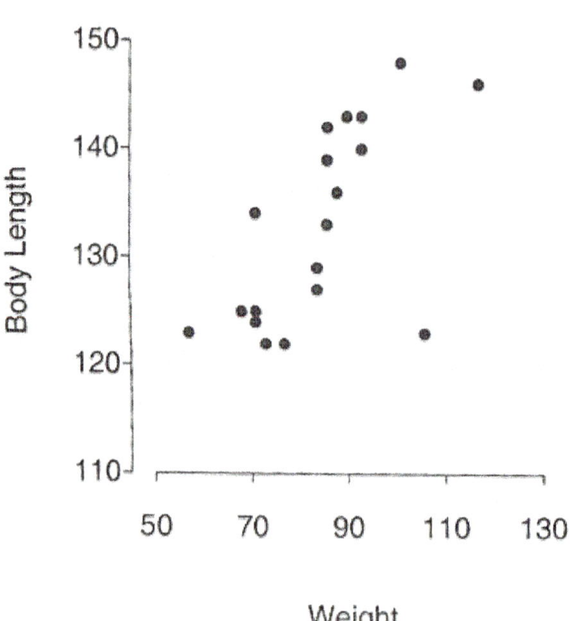

(b)

```
Correlations: bodyleng, weight


Pearson correlation of bodyleng and weight = 0.649
```

3.61 (a) The scatter diagram is shown below.

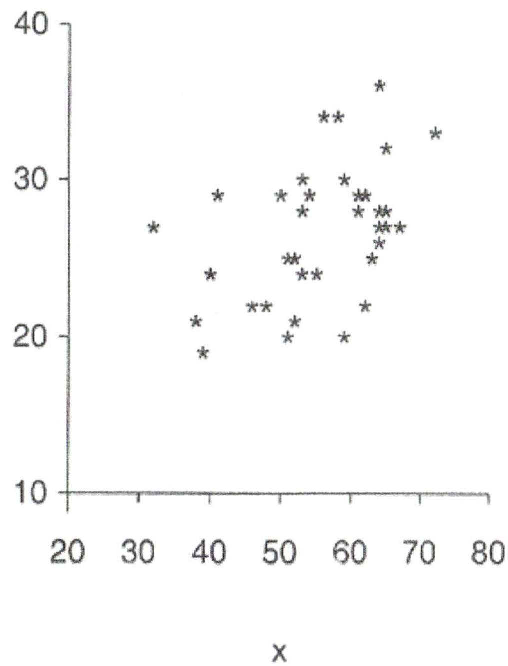

(b) With $n = 36$, we calculate

$$\sum x = 2006 \quad , \quad \sum y = 961$$
$$\sum x^2 = 114{,}950 \qquad \sum xy = 54{,}166 \qquad \sum y^2 = 26{,}281$$

so

$$S_{xx} = \sum x^2 - \frac{\left(\sum x\right)^2}{n} = 114{,}950 - \frac{(2006)^2}{36} = 3{,}171.22$$
$$S_{yy} = \sum y^2 - \frac{\left(\sum y\right)^2}{n} = 26{,}281 - \frac{(961)^2}{36} = 627.64$$
$$S_{xy} = \sum xy - \frac{\left(\sum x\right)\left(\sum y\right)}{n} = 54{,}166 - \frac{(2006)(961)}{36} = 616.94$$

Consequently,

$$r = \frac{S_{xy}}{\sqrt{S_{xx}}\sqrt{S_{yy}}} = \frac{616.94}{\sqrt{3{,}171.22}\sqrt{627.64}} = 0.437.$$

Chapter 4

PROBABILITY

4.1 (a) (iv), (v) (e) (v)
(b) (ii), (v) (f) (i)
(c) (vi) (g) (iii), (v)
(d) (vi)

4.3 (a) (ii) (b) (iii) (c) (i)

4.5 (a) {0, 1}

(b) {0, 1,..., 344}

(c) $\{t : 90 < t < 425.4\}$

4.7 (a) Let us identify Bob, John, Linda, and Sue by their initials B, J, L, and S, respectively. We make a tree diagram:

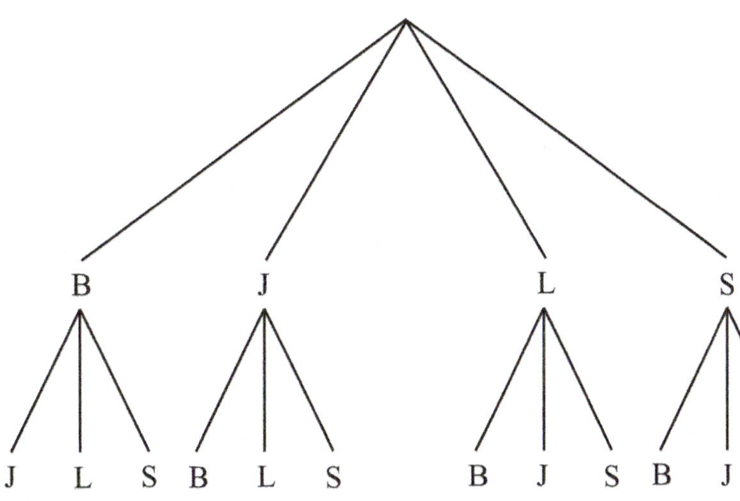

S = {BJ, BL, BS, JB, JL, JS, LB, LJ, LS, SB, SJ, SL}

(b) A = {LB, LJ, LS}, B = {JL, LJ, JS, SJ, LS, SL}

4.9 $P(e_1)+P(e_2)+P(e_3)=0.3+0.4+0.2=0.9$. Since $P(S)=1$, we must have $P(e_4)=1-0.9=0.1$.

4.11 (a) yes
(b) no, because the sum of probabilities is less than 1
(c) yes

4.13 Denote May by e_1 and so on. Because $1+3+6+10=20$, we have

$$P(e_1)=\frac{1}{20},\ P(e_2)=\frac{3}{20},\ P(e_3)=\frac{6}{20},\ P(e_4)=\frac{10}{20}$$

so that $P(A)=P(e_1)+P(e_2)=\frac{4}{20}=0.2$.

4.15 The relative frequencies are based on a very large number of cases and will therefore be very good approximations to the probabilities. Since

$$P(\text{weekday}) + P(\text{weekend}) = 1$$

and P(weekend) is approximately 0.286, the probability of a weekday birth is approximately 1 - 0.286 = 0.714

4.17 (a) The tree diagram is

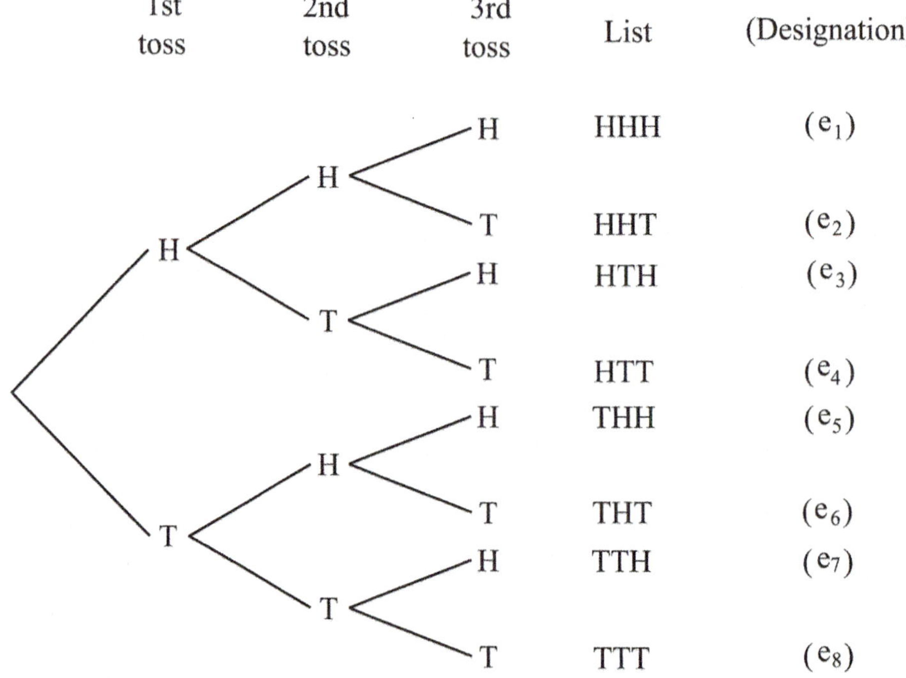

$S = \{e_1, e_2, ..., e_8\}$

(b) Assuming the coins are all fair, all the elementary outcomes are equally likely.

$$P(e_1) = P(e_2) = \cdots = P(e_8) = \frac{1}{8}$$

(c) [Exactly one head] $= \{e_4, e_6, e_7\}$, its probability is $\frac{3}{8}$.

4.19 (a) Let e_1, e_2, and e_3 denote the outcomes of getting a ticket numbered 1, 2, and 3, respectively. Then $S = \{e_1, e_2, e_3\}$. Since all 8 tickets are equally likely to be drawn, and there are 2 tickets with number 1, we have $P(e_1) = \frac{2}{8}$. Likewise, $P(e_2) = \frac{3}{8}$ and $P(e_3) = \frac{3}{8}$.

(b) [Odd-numbered ticket drawn] $= \{e_1, e_3\}$, so the probability is

$$P(e_1) + P(e_3) = \frac{2}{8} + \frac{3}{8} = \frac{5}{8} = 0.625.$$

4.21 (a) Each elementary outcome is a pair of numbers, the first corresponds to the white die and the second to the colored die.

$$A = \{(1, 5), (2, 4), (3, 3), (4, 2), (5, 1)\}$$
$$B = \{(1, 6), (2, 5), (3, 4), (4, 3), (5, 2), (6, 1)\}$$
$$C = \{(2, 6), (4, 6), (6, 6), (1, 5), (3, 5), (5, 5),$$
$$(2, 4), (4, 4), (6, 4), (1, 3), (3, 3), (5, 3),$$
$$(2, 2), (4, 2), (6, 2), (1, 1), (3, 1), (5, 1)\}$$
$$D = \{(1, 1), (2, 2), (3, 3), (4, 4), (5, 5), (6, 6)\}.$$

(b) Probability $\frac{1}{36}$ for each elementary outcome.

(c) $P(A) = \frac{5}{36}$, $P(B) = \frac{6}{36} = \frac{1}{6}$, $P(C) = \frac{18}{36} = \frac{1}{2}$, $P(D) = \frac{6}{36} = \frac{1}{6}$

4.23 (a) The tree diagram is

Answer to Q_1	Answer to Q_2	List
T	T	TT
	F	TF
	I	TI
F	T	FT
	F	FF
	I	FI
I	T	IT
	F	IF
	I	II

So, S= {TT, TF, TI, FT, FF, FI, IT, IF, II} .

(b) Because the student selects the answers at random, the 9 elementary outcomes in S are all equally likely, each has a probability $\frac{1}{9}$. Let us suppose that the correct answers are T for Q_1 and T for Q_2. Then, the event "one correct answer" has the composition {TF, TI, FT, IT}, so that

$$P(\text{one correct answer}) = \frac{4}{9}.$$

Note: Whatever be the correct answers for Q_1 and Q_2, there will be four cases in which one marked answer will match and one will not match.

4.25 (a) The 15 persons are equally likely to be selected. Among them there is only one of blood group AB, so that $P[\text{AB}] = \frac{1}{15}$.

(b) The number of persons of blood group either A or B is $5+6=11$, so that the required probability is $\frac{11}{15}$.

(c) $P[\text{not O}] = \frac{5+6+1}{15} = \frac{12}{15}$.

4.27 $S = \{\text{N, YN, YYN, YYYN, YYYYN, YYYYY}\}$

4.29 (a) Letting c, b, and v denote "compliance", "borderline case", and "violation", respectively,

$$S = \{c_1, c_2, \ldots, c_9, b_1, b_2, b_3, v_1, v_2\}.$$

(b) The 14 elementary outcomes are equally likely, and two of them, namely v_1 and v_2, constitute the event that a violator is detected. The probability is $2/14 = 0.143$.

4.31 (a) The successive days cannot be considered as independent trials. The rate on one day is the same as, or is very close to, the rate on the next day. The results for successive days are not independent. There may also be a trend in rates over the year.

(b) Cars brought in with other problems are more likely to have an emission problem. For instance, there would be too many old cars in this sample.

4.33 (a) Since the gift certificates are assigned at random, all 6 elementary outcomes are equally likely, so that each has probability 1/6.

(b) $P(A) = \frac{3}{6} = \frac{1}{2}$, $\quad P(B) = \frac{2}{6} = \frac{1}{3}$.

4.35 (a) The Venn diagram is

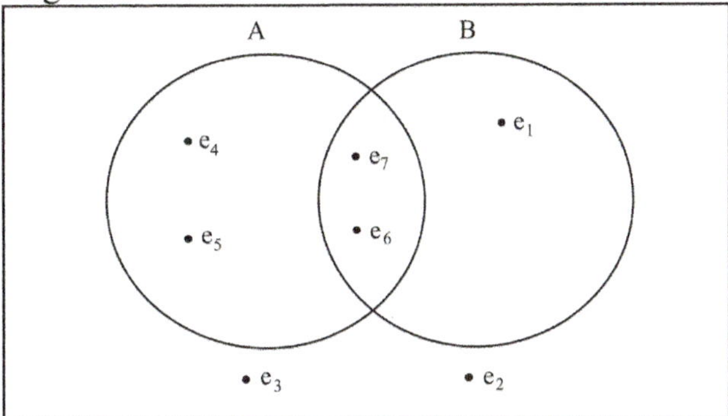

Figure 4.1: Venn Diagram for Exercise 4.35(a)

(b) (i) $AB = \{e_6, e_7\}$

(ii) $\overline{B} = \{e_2, e_3, e_4, e_5\}$

(iii) $A\overline{B} = \{e_4, e_5\}$

(iv) $A \cup B = \{e_1, e_4, e_5, e_6, e_7\}$

(b) (i) $\overline{B} = \{e_1, e_4, e_5, e_8\}$, $P(\overline{B}) = 0.07 + 0.11 + 0.15 + 0.15 = 0.48$

(ii) $BC = \{e_6\}$, $P(BC) = 0.15$

(iii) $A \cup C = \{e_1, e_2, e_5, e_6, e_7, e_8\}$,
$P(A \cup C) = 0.07 + 0.11 + 0.15 + 0.15 + 0.15 + 0.15 = 0.78$

(iv) $\overline{A} \cup C = \{e_3, e_4, e_6, e_8\}$,
$P(\overline{A} \cup C) = 0.11 + 0.11 + 0.15 + 0.15 = 0.70$

4.37 (a) $\overline{C} = \{e_1, e_2, e_3, e_4, e_5, e_7\}$, $P(\overline{C}) = 0.07 + 0.11 + 0.11 + 0.11 + 0.15 + 0.15 = 0.70$

(b) $AB = \{e_2, e_6, e_7\}$, $P(AB) = 0.11 + 0.15 + 0.15 = 0.41$

(c) $A\overline{B} = \{e_1, e_5\}$, $P(A\overline{B}) = 0.07 + 0.15 = 0.22$

(d) $\overline{A}\overline{C} = \{e_3, e_4\}$, $P(\overline{A}\overline{C}) = 0.11 + 0.11 = 0.22$

4.39 (a) Denote by e_1, e_2, e_3, and e_4 the elementary outcomes that the person hired is candidate number 1, 2, 3, and 4, respectively. The Venn diagram is given in Figure 4.2.

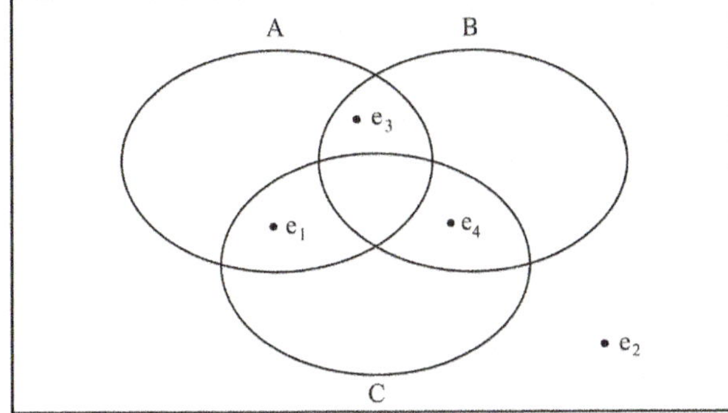

Figure 4.2: Venn diagram for Exercise 4.39(a)

(b) $A \cup B = \{e_1, e_3, e_4\}$ and $AB = \{e_3\}$

4.41 (a) $P(A) = 0.04 + 0.2 + 0.06 = 0.30$ and $P(B) = 0.04 + 0.2 + 0.06 + 0.06 = 0.36$
Also, since $AB = \{e_5, e_8\}$, $P(AB) = 0.2 + 0.06 = 0.26$.

(b) $P(A \cup B) = P(A) + P(B) - P(AB) = 0.30 + 0.36 - 0.26 = 0.40$

(c) Since $A \cup B = \{e_1, e_5, e_8, e_2, e_9\}$,
$P(A \cup B) = 0.04 + 0.2 + 0.06 + 0.04 + 0.06 = 0.40$

(d) $P(\overline{B}) = 1 - P(B) = 1 - 0.36 = 0.64$. Alternatively, since $\overline{B} = \{e_1, e_3, e_4, e_6, e_7\}$,
$P(\overline{B}) = 0.04 + 0.2 + 0.2 + 0.1 + 0.1 = 0.64$.

4.43 (a) The specified probabilities are entered in the table and underlined. The other entries are obtained in part (b).

	B	$\overline{B}$	
A	<u>0.18</u>	0.18	<u>0.36</u>
$\overline{A}$	0.32	0.32	0.64
	<u>0.50</u>	0.50	1.00

(b) Since $A = AB \cup A\overline{B}$, a union of mutually exclusive events, we have $P(A) = P(AB) + P(A\overline{B})$. Hence, for the given information, $0.36 = 0.18 + P(A\overline{B})$. Thus, solving for $P(A\overline{B})$ yields:

$$P(A\overline{B}) = 0.36 - 0.18 = 0.18.$$

Similarly,
$P(\overline{A}B) = 0.64 - 0.32 = 0.32$, $P(\overline{A}\overline{B}) = 1 - (0.18 + 0.18 + 0.32) = 0.32$

4.45 (a) The completed probability table is given below

	B	$\overline{B}$	
A	0.14	0.23	0.37
$\overline{A}$	0.36	0.27	0.63
	0.50	0.50	1.00

(b) $P(A\overline{B}) = 0.37 - 0.14 = 0.23$
(c) $P(A \cup B) = P(A) + P(B) - P(AB) = 0.37 + 0.50 - 0.14 = 0.73$
(d) $P(A\overline{B} \cup \overline{A}B) = P(A\overline{B}) + P(\overline{A}B)$ (union of incompatible events)
$= 0.23 + 0.36 = 0.59$

4.47 (a) $P(\overline{A}) = 0.4 + 0.25 = 0.65$
(b) $P(A\overline{B}) = 0.15$
(c) $P(A\overline{B}) + P(\overline{A}B) = 0.15 + 0.4 = 0.55$

4.49 Denoting 'violation' by V and 'compliance' by C, the classification of the 18 restaurants is shown in the following table:

		Safety V	C	Total
Sanitary	V	4	3	7
	C	4	7	11
		8	10	18

$$P(CC) = \frac{7}{18} = 0.389$$

4.51 (a) $P(A) = 0.08 + 0.02 + 0.20 + 0.10 = 0.40$

$P(B) = 0.15 + 0.10 + 0.08 + 0.02 = 0.35$

$P(BC) = 0.15 + 0.08 = 0.23$

$P(ABC) = 0.08$

(b) (i) Light case and above 40.

$$P(\bar{A}\bar{B}) = 0.15 + 0.20 = 0.35$$

(ii) Either a light case or the parents are not diabetic or both.

$$P(\bar{A} \cup \bar{C}) = 0.15 + 0.10 + 0.15 + 0.20 + 0.02 + 0.10 = 0.72$$

(iii) A light case, age is below 40 and parents are not diabetic.

$$P(\bar{A}B\bar{C}) = 0.10\,.$$

4.53 (a) With the stated numbers identifying the gift boxes, the list is:

(1,1), (1,2), (1,3), (1,4), (1,5)

(2,1), (2,2), (2,3), (2,4), (2,5)

(3,1), (3,2), (3,3), (3,4), (3,5)

(4,1), (4,2), (4,3), (4,4), (4,5)

(5,1), (5,2), (5,3), (5,4), (5,5)

The 25 elementary outcomes are equally likely, so each has the probability $\frac{1}{25}$.

(b) $A = \{(2,1), (2,2), (2,3), (2,4), (2,5), (3,1), (3,2), (3,3), (3,4), (3,5), (1,2), (4,2), (5,2), (1,3), (4,3), (5,3)\}$,

So, $P(A) = \frac{16}{25}$.

$B = \{(3,1), (3,2), (3,3), (3,4), (3,5), (4,1), (4,2), (4,3), (4,4), (4,5), (5,1), (5,2), (5,3), (5,4), (5,5), (1,3), (1,4), (1,5), (2,3), (2,4), (2,5)\}$

So, $P(B) = \frac{21}{25}$.

$AB = \{(2,3), (2,4), (2,5), (1,3), (3,1), (3,2), (3,3), (3,4), (3,5), (4,2), (4,3), (5,2), (5,3)\}$,

So, $P(AB) = \frac{13}{25}$.

4.55 It is reasonable to expect $P(A \mid B) > P(A)$, since the set of luxury car owners would include a large percentage of the set of lawyers. By this reasoning, A and B are not independent for otherwise we would have $P(A \mid B) = P(A)$.

4.57 Observe that $P(B \mid A) = \frac{P(AB)}{P(A)} = \frac{0.001}{0.101} = 0.0099$ and $P(B) = 0.05 + 0.001 = 0.051$.

Since $P(B) \neq P(B \mid A)$, the events A and B are not independent.

4.59 (a) $P(\bar{A}) = 1 - P(A) = 1 - 0.4 = 0.6$

(b) $P(AB) = P(B)P(A \mid B) = 0.25 \times 0.7 = 0.175$

(c) $P(A \cup B) = P(A) + P(B) - P(AB) = 0.4 + 0.25 - 0.175 = 0.475$

4.61

2	Green
3	Red
5	

$\to 2$, without replacement

(a) We denote G for green, R for red and attach subscripts to identify the order of the draws. Since the event A, a green ball appears in the first draw, has nothing to do with the second draw, we identify $A = G_1$ so

$$P(A) = P(G_1) = \frac{2}{5} = 0.4$$

The event $B = G_2$ is the union of G_1G_2 and R_1G_2.

$$P(G_1G_2) = P(G_1)P(G_2 \mid G_1) = \frac{2}{5} \times \frac{1}{4} = \frac{2}{20}$$

$$P(R_1G_2) = P(R_1)P(G_2 \mid R_1) = \frac{3}{5} \times \frac{2}{4} = \frac{6}{20}$$

Hence, $P(B) = P(G_2) = \frac{2}{20} + \frac{6}{20} = \frac{8}{20} = 0.4$

(b) $P(AB) = P(G_1G_2) = \frac{2}{20} = 0.1$. On the other hand, $P(A)P(B) = 0.4 \times 0.4 = 0.16$, and this is different from $P(AB)$. Therefore, A and B are not independent.

(c) Now, we redo parts (a) and (b) *with* replacement:

Part (a): $P(A) = P(G_1) = \frac{2}{5} = 0.4$

$P(G_1G_2) = P(G_1)P(G_2 \mid G_1) = \frac{2}{5} \times \frac{2}{5} = \frac{4}{25}$

$P(R_1G_2) = P(R_1)P(G_2 \mid R_1) = \frac{3}{5} \times \frac{2}{5} = \frac{6}{25}$

Adding, we get $P(B) = P(G_2) = \frac{4}{25} + \frac{6}{25} = \frac{10}{25} = 0.4$

Part (b): $P(AB) = P(G_1G_2) = \frac{4}{25} = 0.16$

$P(A)P(B) = 0.4 \times 0.4 = 0.16 = P(AB)$.

Since these probabilities are equal, the events are independent.

4.63 We are given the following information:

$$\text{P(making a claim)} = \frac{2{,}073}{12{,}299} \approx 0.1686$$

$$\text{P(making a claim} \mid \text{age} < 25) = \frac{1{,}032}{5{,}192} \approx 0.1988$$

We now can create the following table:

	Age < 25	Age ≥ 25	
Claim	0.0839	0.0847	0.1686
No claim	0.3382	0.4932	0.8314
	0.4221	0.5779	

(a) P(make claim) = 0.1686 (This was given.)
(b) P(claim and age < 25) = 0.0839
(c) The data does not suggest the need to do so because the probabilities that someone is < 25 and that someone is > 25 file a claim are practically the same – see the first row in the table.

4.65 (a) If independent, $P(AB) = P(A)P(B) = 0.6 \times 0.22 = 0.132$ so
$P(A \cup B) = P(A) + P(B) - P(AB) = 0.6 + 0.22 - 0.132 = 0.688$

(b) If mutually exclusive, $P(AB) = 0$ so

$$P(A \cup B) = P(A) + P(B) = 0.6 + 0.22 = 0.82$$

(c) If A and B are mutually exclusive, we have $A\bar{B} = A$ so

$$P(A \mid \bar{B}) = \frac{P(A\bar{B})}{P(\bar{B})} = \frac{0.6}{0.78} = 0.769.$$

4.67 The classification of the 20 rats is shown in the following table:

	Infected (I)	Not infected (N)	Total
Male (M)	7	5	12
Female (F)	2	6	8
	9	11	20

(a) There are 9 infected rats of which 2 are females. Therefore, $P(F \mid I) = \frac{2}{9}$.

Alternatively, we can use the definition of conditional probability

$$P(F \mid I) = \frac{P(FI)}{P(I)} = \frac{2/20}{9/20} = \frac{2}{9}.$$

(b) There are 12 males of which 7 are infected, so $P(I \mid M) = \frac{7}{12}$.

(c) $P(I) = \frac{9}{20}, \quad P(M) = \frac{12}{20}, \quad P(IM) = \frac{7}{20} = 0.35$

$P(I)P(M) = \frac{9}{20} \times \frac{12}{20} = 0.27$

Since $P(IM) \neq P(I)P(M)$, the events are not independent.

4.69 (a) BC, $P(BC) = 0$ since B and C are incompatible.

(b) $A \cup B$, $P(A \cup B) = P(A) + P(B) - P(AB)$

Now, by independence, $P(AB) = P(A)P(B) = 0.7 \times 0.2 = 0.14$

Hence, $P(A \cup B) = 0.7 + 0.2 - 0.14 = 0.76$

(c) $\bar{B}$, $P(\bar{B}) = 1 - P(B) = 1 - 0.2 = 0.8$

(d) ABC, $P(ABC) = 0$ since B and C are incompatible and so are AB and C.

4.71 Denote the events by:

S = the cooling system functions
S_1 = the primary unit functions
F_1 = the primary unit fails
S_2 = the back-up unit functions
F_2 = the back up unit fails

Then S can be expressed as the union of two incompatible events: $S = S_1 \cup (F_1 S_2)$ As such, $P(S) = P(S_1) + P(F_1 S_2)$. Next, observe that $P(S_1) = .999$ and

$$P(F_1 S_2) = P(F_1)P(S_2 \mid F_1) = 0.001 \times 0.910 = 0.00091.$$

Therefore, $P(S) = 0.999 + 0.00091 = 0.99991$.

4.73 (a) Denoting the success and failure in each test by S and F, respectively, the sample space is conveniently listed with the tree diagram below.

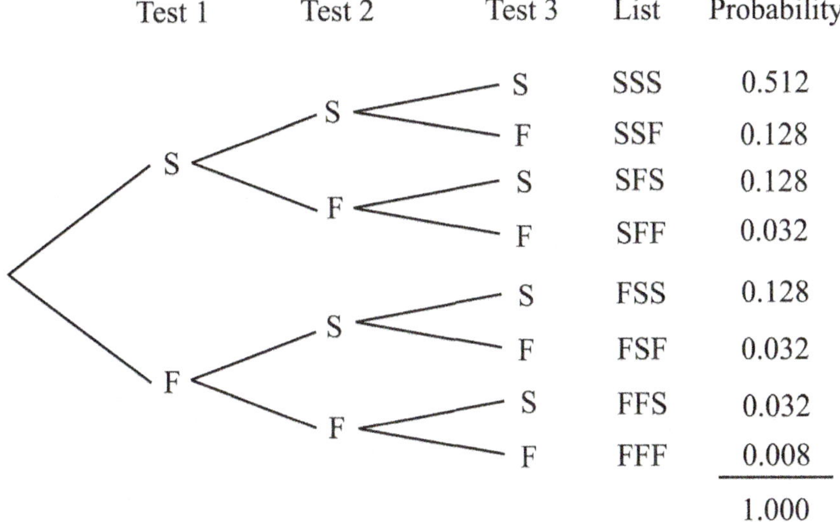

To calculate the probabilities we note that $P(S) = 0.8$ and $P(F) = 1 - 0.8 = 0.2$. Since the tests are independent, we calculate

$$P(SSS) = P(S)P(S)P(S) = 0.8 \times 0.8 \times 0.8 = 0.512$$
$$P(SSF) = P(S)P(S)P(F) = 0.8 \times 0.8 \times 0.2 = 0.128, \text{ etc.}$$

The results are shown in the column of probability in the above figure.

(b) P(at least two successes) = P(2 successes) + P(3 successes) $= (0.128 + 0.128 + 0.128) + 0.512 = 0.896$.

4.75 Denote the events

$$S = \text{strep-throat} \quad \text{and} \quad A = \text{allergy}$$

We are given that $P(S) = 0.25$, $P(A) = 0.4$, and $P(SA) = 0.1$.

(a) $P(S \cup A) = P(S) + P(A) - P(SA) = 0.25 + 0.4 - 0.1 = 0.55$

(b) $P(S)P(A) = 0.25 \times 0.4 = 0.10 = P(SA)$, so the events are independent.

4.77

$$P(B|A) = \frac{P(A|B)P(B)}{P(A|B)P(B) + P(A|\overline{B})P(\overline{B})}$$
$$= \frac{(0.995)(0.014)}{(0.995)(0.014) + (0.03)(0.986)} = 0.320$$

He has about a 32% chance of having the disease given that he tested positive.
4.78 (a) By the rule of total probability,

$$P(\text{Incomplete}) = P(\text{Incomplete}|\text{Carol did repair}) \cdot P(\text{Carol did repair})$$
$$+ P(\text{Incomplete}|\text{Karl did repair}) \cdot P(\text{Karl did repair})$$
$$= (0.04)(0.60) + (0.06)(0.40) = 0.048$$

(a) By Bayes' Formula,

$$P(\text{Karl did repair}|\text{ Repair incomplete})$$
$$= \frac{P(\text{Incomplete}|\text{Karl did repair}) \cdot P(\text{Karl did repair})}{P(\text{Incomplete}|\text{Karl did repair}) \cdot P(\text{Karl did repair}) + P(\text{Incomplete}|\text{Carol did repair}) \cdot P(\text{Carol did repair})}$$
$$= \frac{(0.06)(0.40)}{(0.06)(0.40) + (0.04)(0.60)} = 0.50$$

4.79 We use the symbols M for male, F for female, U for unemployed and E for employed.

(a) $P(M) = 0.6$
$P(U \mid M) = 0.051$
$P(U \mid F) = 0.043$

(b) $P(UM) = 0.051 \times 0.6 = 0.0306$ and $P(UF) = 0.043 \times (1 - 0.6) = 0.0172$
Adding these we obtain $P(U) = 0.0306 + 0.0172 = 0.0478$, so the overall rate of unemployment is 4.8%.

(c) To find $P(F \mid U)$, we use the results $P(U) = 0.0478$ and $P(UF) = 0.0172$ and obtain $P(F \mid U) = \frac{P(FU)}{P(U)} = \frac{0.0172}{0.0478} = 0.360$.

4.81 Let A = 1st serve in bounds, so that $\overline{A}$= 1st serve not in bounds. Also, let B = wins point, so that $\overline{B}$= does not win point.

We have the following probabilities:
P(A) = 0.61, P($\overline{A}$)=1-0.61 = 0.39, P(B|A) = 0.62, and $P(B \mid \overline{A}) = 0.54$.

(a) $P(B) = P(B \mid A)P(A) + P(B \mid \bar{A})P(\bar{A}) = (0.62)(0.61) + (0.54)(0.39) = 0.5888$

(b) $P(A \mid B) = \dfrac{P(B \mid A)P(A)}{0.5888} = \dfrac{(0.62)(0.61)}{0.5888} = 0.6423$

4.83 (a) Identify the events in the same manner as in Example 22. Observe that

$$P(B_2) = 0.30,\ P(B_3) = 0.15 .$$

So, $P(B_1) = 1 - 0.30 - 0.15 = 0.55$.

The conditional probabilities are the same, so

$$P(A) = (0.04)(0.55) + (0.07)(0.30) + (0.10)(0.15) = 0.058$$

(b) $P(B_2 \mid A) = \dfrac{(0.07)(0.30)}{0.058} \approx 0.3621$

So, Supplier 2 produces over 36% of the defective ones.

4.85 (a) The number of possible selections of 4 persons out of 10 persons is

$$\binom{10}{4} = \frac{10 \times 9 \times 8 \times 7}{4 \times 3 \times 2 \times 1} = 210$$

(b) The number of possible selections of 2 men out of 6 men is

$$\binom{6}{2} = \frac{6 \times 5}{2 \times 1} = 15$$

and the number of possible selections of 2 women out of 4 women is

$$\binom{4}{2} = \frac{4 \times 3}{2 \times 1} = 6$$

Since the men can be selected in 15 ways, and for each selection of men, there are 6 ways the women can be selected, the number of possible selections of two men and two women is $15 \times 6 = 90$.

4.87 (a) The number of possible selections of 5 children out of 11 is $\binom{11}{5} = 462$.

(b) The number of selections of 2 out of the 4 young males is $\binom{4}{2} = 6$.

The number of selections of 3 out of the 7 young females is $\binom{7}{3} = 35$.

Each of the 6 choices of the young males can accompany each of the 35 choices of the young females, so the number of selections of 2 males and 3 female students is $\binom{4}{2} \times \binom{7}{3} = 6 \times 35 = 210$.

4.89 The number of possible samples of 5 jurors out of 17 is $\binom{17}{5} = 6188$. Under random sampling, these 6188 possible samples are equally likely. The number of possible samples where all 5 jurors selected are males is $\binom{10}{5} = 252$. The jury selection shows discrimination since

$$P(\text{no female members}) = \frac{\binom{10}{5}}{\binom{17}{5}} = \frac{252}{6188} = 0.041.$$

4.91 The batch contains 4 defective and 16 good alternators. The number of possible samples of size 3 is $\binom{20}{3} = 1140$ and these are equally likely.

(a) $P(A) = \frac{\binom{16}{3}}{\binom{20}{3}} = \frac{560}{1140} = 0.491$ (b) $P(B) = \frac{\binom{4}{2} \times \binom{16}{1}}{\binom{20}{3}} = \frac{6 \times 16}{1140} = 0.084$

4.93 No. The states with fewer seniors get more representation in this process than what the random selection would permit.

4.95 (a) The number of possible selections of 3 plots out of 9 is $\binom{9}{3} = 84$, and the 84 selections are equally likely. One row can be chosen in $\binom{3}{1} = 3$ ways, and within that row, 3 plots can be chosen in $\binom{3}{3} = 1$ way. Therefore, the number of choices such that the three plots are in the same row is $\binom{3}{1} \times \binom{3}{3} = 3 \times 1 = 3$. Hence, the required probability $= \frac{3}{84} = 0.036$.

(b) The number of ways one plot can be selected from row 1 is $\binom{3}{1} = 3$. Likewise, a plot can be selected from row 2 in 3 ways, and from row 3 in 3 ways. The number of possible selections of 3 plots, one in each row, is $3 \times 3 \times 3 = 27$, so the required probability $= \dfrac{27}{84} = 0.321$.

4.97

Row A			Row B		
4	bushy		6	bushy	
4	lean	→2	3	lean	→2
8			9		

(a) There are 8 trees in row A from which 2 trees can be selected in $\binom{8}{2} = 28$ ways. Of these 28 equally likely selections, there are $\binom{4}{2} = 6$ selections in which both trees are bushy. Therefore,

$$P[\text{2 bushy trees selected in row A}] = \frac{6}{28}.$$

Similarly,

$$P[\text{2 lean trees selected in row B}] = \frac{\binom{3}{2}}{\binom{9}{2}} = \frac{3}{36}.$$

By independence of the two selections, the required probability is therefore

$$= \frac{6}{28} \times \frac{3}{36} = 0.018.$$

(b) Let A_0, A_1, and A_2 respectively denote the events of getting exactly 0, 1, or 2 bushy trees in row A, and let B_0, B_1, B_2 denote the corresponding events for row B. Then

[Exactly 2 bushy] $= A_2B_0 \cup A_1B_1 \cup A_0B_2$ (union of mutually exclusive events)

$$P(A_2B_0) = P(A_2)P(B_0) = \frac{\binom{4}{2}}{\binom{8}{2}} \times \frac{\binom{3}{2}}{\binom{9}{2}} = \frac{6}{28} \times \frac{3}{36} = 0.018 \quad \text{(see part (a))}$$

$$P(A_1B_1) = P(A_1)P(B_1) = \frac{\binom{4}{1}\binom{4}{1}}{\binom{8}{2}} \times \frac{\binom{6}{1}\binom{3}{1}}{\binom{9}{2}} = \frac{16}{28} \times \frac{18}{36} = 0.286$$

$$P(A_0B_2) = P(A_0)P(B_2) = \frac{\binom{4}{2}}{\binom{8}{2}} \times \frac{\binom{6}{2}}{\binom{9}{2}} = \frac{6}{28} \times \frac{15}{36} = 0.089$$

Adding these probabilities we get P[exactly 2 bushy] $= 0.393$.

4.99

5	below thirty	
6	over thirty	$\rightarrow 4$ randomly selected
11		

(a) The number of possible choices of 4 persons out of 11 is $\binom{11}{4} = 330$.

(b) The number of choices of 3 persons below thirty and 1 over thirty is

$$\binom{5}{3} \times \binom{6}{1} = 10 \times 6 = 60.$$

So, the required probability $= \frac{60}{330} = 0.182$.

4.101

6	yellow	
5	red	$\rightarrow 4$
11		

The number of possible samples of 4 bulbs out of 11 is $\binom{11}{4} = 330$, and all choices are equally likely.

(a) The number of ways 2 red and 2 yellow bulbs can be selected is

$$\binom{5}{2} \times \binom{6}{2} = 10 \times 15 = 150$$

So, P[exactly 2 red] $= \frac{150}{330} = 0.455$.

(b) We calculate

$$P[2 \text{ red}] = \frac{150}{330} \text{ (done in part (a))}$$

$$P[3 \text{ red}] = \frac{\binom{5}{3} \times \binom{6}{1}}{330} = \frac{10 \times 6}{330} = \frac{60}{330}$$

$$P[4 \text{ red}] = \frac{\binom{5}{4}}{330} = \frac{5}{330}$$

Adding these probabilities, we obtain

$$P[\text{at least 2 red}] = \frac{150 + 60 + 5}{330} = \frac{215}{330} = 0.652$$

(c) $$P[\text{all 4 red}] = \frac{\binom{5}{4}}{330} = \frac{5}{330}$$

$$P[\text{all 4 yellow}] = \frac{\binom{6}{4}}{330} = \frac{15}{330}$$

Adding these probabilities, we obtain

$$P[\text{all 4 of the same color}] = \frac{20}{330} = 0.061.$$

4.103 (a) $S = \{1, 2, ..., 24\}$

(b) $S = \{p : p > 0\}$, p is tire pressure in psi

(c) $S = \{0, 1, 2, ..., 50\}$

(d) $S = \{t : t \geq 0\}$, t is time in days.

4.105 (a) $A = \{23, 24\}$

(b) $A = \{p : 0 < p \leq 28\}$

(c) Since $0.25 \times 50 = 12.5$, $A = \{0, 1, ..., 12\}$

(d) $A = \{t : 0 \leq t < 500.5\}$

4.107 $S = \{p : 0 \leq p < 100\}$ where $p =$ percentage of alcohol in blood.

$A = \{p : .10 < p < 100\}$

4.109 The sample space is listed by means of a tree diagram which is given in Figure 4.4.

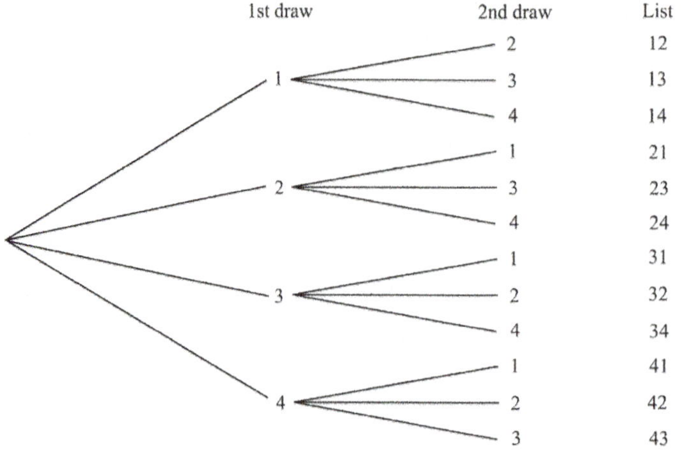

Figure 4.4: Sample space for Exercise 4.109

(a) S consists of 12 elementary outcomes which are equally likely because of random selection. Of the 12 elementary outcomes, 6 correspond to even numbers so

$$P[\text{even number}] = \frac{6}{12} = 0.5$$

(b) There are 9 elementary outcomes where the number is larger than 20, so the required probability is $\frac{9}{12} = 0.75$.

(c) There are 2 elementary outcomes, namely {23, 24}, for which the number is between 22 and 30, so the required probability is $\frac{2}{12} = 0.167$.

4.111 Consider the plot selected from each row to be assigned to variety 'a'. We list the sample space by drawing a tree diagram which is given in the figure below.

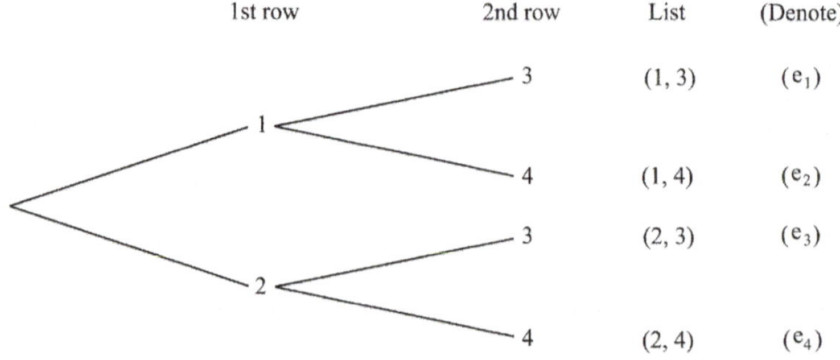

$S = \{e_1, e_2, e_3, e_4\}$ and the elements are equally likely.

[same column] $= \{(1, 3), (2, 4)\}$, Probability $= \frac{2}{4} = \frac{1}{2}$.

4.113 (a) No. Heavy air traffic occurs on Mondays and Fridays.
(b) Yes. The temperature has nothing to do with the day of the week.
(c) No. Ozone lower on weekends when not as much automobile exhaust.
(d) No. December almost always has highest sales.

4.115 (a) Either a faulty transmission or faulty brakes.

(b) Transmission, brakes and exhaust system all faulty.

(c) No faults with the transmission, brakes or the exhaust system.

(d) Either the transmission is not faulty or the brakes are not faulty.

4.117 Because $P(AB) \leq P(A) \leq P(A \cup B)$, we have $P(A) = 0.30$, $P(AB) = 0.1$, and $P(A \cup B) = 0.50$.

4.119 (a) ABC (b) $A \cup B \cup C$ (c) $AB\overline{C}$ (d) $\overline{A}B\overline{C}$

4.121 (a) Student is good at answering both essay and T/F questions.

$$P(BC) = 0.05 + 0.20 = 0.25 \quad \text{(see Figure 4.6 above.)}$$

(b) Student is good at answering at least one of essay or T/F type questions.

$$P(B \cup C) = P(B) + P(C) - P(BC) \qquad \text{(Addition law)}$$

$$P(B) = 0.05 + 0.10 + 0.20 + 0.15 = 0.50$$

$$P(C) = 0.08 + 0.05 + 0.20 + 0.18 = 0.51$$

$$P(BC) = 0.25$$

Therefore, $P(B \cup C) = 0.50 + 0.51 - 0.25 = 0.76$
Alternatively,

$$P(B \cup C) = P(B) + P(\overline{B}C) \qquad \text{(See Figure 4.6 above.)}$$

$$= 0.50 + 0.26 = 0.76$$

(c) Student is good at answering essay questions, but not T/F questions.

$$P(B\overline{C}) = 0.10 + 0.15 = 0.25$$

(d) Student is only good at answering one of the three different types of questions.

$$P(A\overline{B}\overline{C} \cup \overline{A}B\overline{C} \cup \overline{A}\overline{B}C) = 0.14 + 0.15 + 0.18 = 0.47$$

4.123 The probabilities can be determined either by using the Venn diagram or the probability table presented in the solution of Exercise. 4.122

(a) $P(B\overline{C}) = 0.05 + 0.20 = 0.25$
(by summing all probabilities in B but outside of C)

(b) $P(A \cup B) = P(A) + P(B) - P(AB)$ (Addition law)

$= 0.51 + 0.45 - 0.17$

$= 0.79$

(c) P[exactly two of the three events occur] $= P(AB\bar{C}) + P(A\bar{B}C) + P(\bar{A}BC)$

$= 0.05 + 0.21 + 0.08$ (see Figure 4.7)

$= 0.34$.

4.125 $P(A \mid B) = \dfrac{P(AB)}{P(B)} = \dfrac{0.4}{0.5} = 0.8$

$P(A)P(B) = 0.8 \times 0.5 = 0.40 = P(AB)$, so A and B are independent.

4.127 (a) $P(AC) = 0.15$

$P(A)P(C) = 0.6 \times 0.25 = 0.15$

Because $P(AC) = P(A)P(C)$, the events A and C are independent.

(b) $P(A\bar{B}C) = P(AC) = 0.15$ (see the Venn diagram in Exercise 4.126)

$P(A\bar{B}) = 0.25 + 0.15 = 0.40$

$P(C) = 0.15 + 0.1 = 0.25$

$P(A\bar{B})P(C) = 0.40 \times 0.25 = 0.1 \neq P(A\bar{B}C)$, so the events $A\bar{B}$ and C are not independent.

4.129 (a) $P(A \mid \bar{B}) = \dfrac{P(A\bar{B})}{P(\bar{B})}$

From the probability table in Exercise 4.120, we get

$P(A\bar{B}) = 0.08 + 0.14 = 0.22$

$P(\bar{B}) = 0.08 + 0.14 + 0.18 + 0.10 = 0.50$

So, $P(A \mid \bar{B}) = \dfrac{0.22}{0.50} = 0.44$.

(b) $P(B \mid AC) = \dfrac{P(ABC)}{P(AC)} = \dfrac{0.05}{0.13} = 0.3846$

(c) From the probability table, we find $P(A) = 0.37$, $P(C) = 0.51$. $P(AC) = 0.13$.

Since $P(A)P(C) = 0.37 \times 0.51 = 0.1887 \neq P(AC)$, the events A and C are dependent.

4.131 (a) P(BA) = P(A)P(B|A) = (0.40)(0.0002) = 0.00008, or 8 out of 100,000 times.

(b) P(BA) = (0.10)(0.0002) = 0.00002, or 2 out of 100,000 times.

4.133 (a) P(file complaint | 20-29 years old) = $\dfrac{56,689}{43.2\times10^{6}} = 0.0013$.

(b) We have the following table:

	20-29	At least 30
File complaint	(0.0013)(0.139) = 0.0001807	(0.000719)(0.861) = 0.0006187
Does not file complaint	0.1388193	0.860381
	0.139	0.861

Let x = total population. Observe that $0.139x = 43.2$, so that $x = 310.79$ million. So, the number of complaints issued by the "at least 30" age group is 223,311. Thus,

$$\frac{223,311}{310.79\times10^{6}} \approx 0.000719$$

Thus, P(file complaint if at least 30) = 0.0006187.

(c) P(file complaint) = P(file complaint| 20-29)P(20-29) + P(file complaint| at least 30)P(at least 30)
= 0.0001807(0.139) + 0.0006187(0.861) = 0.0005578

So, about 0.06% chance.

(d) $\text{P}(20-29|\text{ files complaint}) = \dfrac{P(\text{file complaint}|\ 20\text{-}29)P(20\text{-}29)}{0.0005578} = 0.0450$.

4.135 (a) & (b) The tree diagram is given below.

Subject 1	Subject 2	List	Probability
F	F	FF	(0.4)(0.4) = 0.16
	D	FD	(0.4)(0.3) = 0.12
	N	FN	(0.4)(0.3) = 0.12
D	F	DF	(0.3)(0.4) = 0.12
	D	DD	(0.3)(0.3) = 0.09
	N	DN	(0.3)(0.3) = 0.09
N	F	NF	(0.3)(0.4) = 0.12
	D	ND	(0.3)(0.3) = 0.09
	N	NN	(0.3)(0.3) = 0.09
			1.00

(i) The elementary outcome probabilities are contained within the above diagram.
(ii) $P(FF)+P(FD)+P(FN)+P(DF)+P(NF)=0.64$
(iii) $P(FF)+P(FD)+P(DF)+P(DD)=0.49$

4.137 There are $\binom{11}{3} = \frac{11 \times 10 \times 9}{3 \times 2 \times 1} = 165$ ways to select three trucks at random. If the company's argument is correct, that there are exactly three noncompliant trucks, then only one selection can consist of three noncompliant vehicles. The probability of selecting all three noncompliant trucks is $\frac{1}{165} = 0.0061$, a small probability. It is quite unlikely a random selection would produce the indicated sample so we can question the veracity of the company's claim.

4.139 Denote $G =$ Good standing, $D =$ Illegal deduction.

11	G	
7	D	$\to 4$
18		

(a) The number of possible selections of 4 returns out of 18 is

$$\binom{18}{4} = 3060, \quad \text{all equally likely.}$$

$$P[\text{Sample has all 4 } G\text{'s}] = \frac{\binom{11}{4}}{\binom{18}{4}} = \frac{330}{3060} = .108 .$$

(b) $P[\text{at least 2 } D\text{'s}] = P[2D's] + P[3D's] + P[4D's]$

$$P[2D's] = \frac{\binom{7}{2}\binom{11}{2}}{\binom{18}{4}} = \frac{21 \times 55}{3060} = \frac{1155}{3060}$$

$$P[3D's] = \frac{\binom{7}{3}\binom{11}{1}}{\binom{18}{4}} = \frac{35 \times 11}{3060} = \frac{385}{3060}$$

$$P[4D's] = \frac{\binom{7}{4}}{\binom{18}{4}} = \frac{35}{3060}$$

Adding these we obtain $P[\text{at least 2 } D\text{'s}] = \frac{1155}{3060} + \frac{385}{3060} + \frac{35}{3060} = \frac{1575}{3060} = 0.515$.

4.141 (a) $P(\text{no common birthday}) = \dfrac{365 \times 364 \times 363}{365 \times 365 \times 365} = 1 \times \dfrac{364}{365} \times \dfrac{363}{365} = 0.992$.

Hence, $P(\text{at least two have the same birthday}) = 1 - 0.992 = 0.008$.

(b) For each person there are 365 possible birthdays. Hence, for N persons, the number of possible birthdays is

$$365 \times 365 \times ... \times 365 \ (N \text{ factors}) = (365)^N .$$

In order that the birthdays of N persons to be all different, the first person can have any of the 365 days, the second person can have any of the remaining $365 - 1 = 364$ days, the third person can have any of the remaining $365 - 2 = 363$ days, ... the Nth person can have any of the remaining $365 - N + 1$ days.

Hence, the number of ways N persons can have all different birthdays is $365 \times 364 \times ... \times (365 - N + 1)$. Therefore,

$$P[\text{No common birthday}] = \frac{365 \times 364 \times \ldots \times (365 - N + 1)}{365^N}$$

Chapter 5

PROBABILITY DISTRIBUTIONS

5.1 (a) Discrete, (b) Continuous, (c) Continuous

(d) Continuous, (e) Discrete

5.3 (a)

Possible choices	x
{1, 3}	2
{1, 5}	4
{1, 6}	5
{1, 7}	6
{3, 5}	2

Possible choices	x
{3, 6}	3
{3, 7}	4
{5, 6}	1
{5, 7}	2
{6, 7}	1

(b) The distinct values of x and the corresponding probabilities are listed in the following table. All 10 choices, listed in part (a), are equally likely so

$$P[X = x] = \frac{\text{No. Choices for which } X = x}{10}$$

x	$P[X = x]$
1	0.2
2	0.3
3	0.1
4	0.2
5	0.1
6	0.1

5.5 (a) The ratings by the judges are given in the figure below.

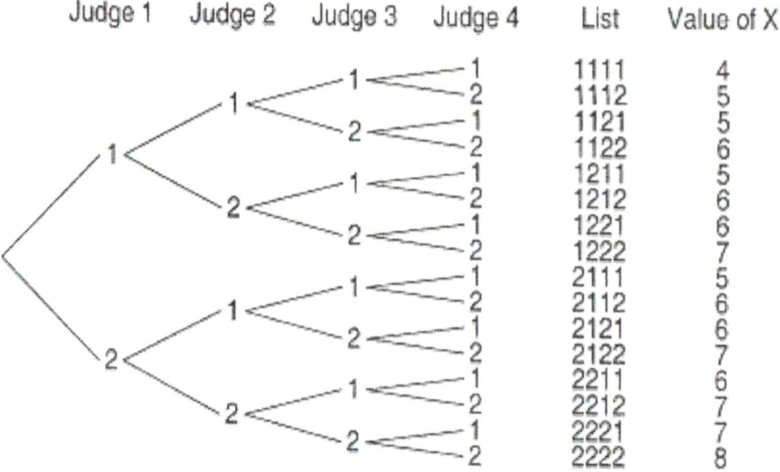

(b) 4, 5, 6, 7, 8

5.7 (a) & (b) The outcomes and the values of X are given on the tree diagram below.

Week 1	Week 2	Week 3	List	Value of X
C	C	C	CCC	0
		B	CCB	1
	B	C	CBC	2
		B	CBB	1
B	C	C	BCC	1
		B	BCB	2
	B	C	BBC	1
		B	BBB	0

5.9 Let X = sum of years of experience of two workers.
There are 10 possible outcomes that give rise to five possible values for this random variable, as described below:

Outcome	Value of X
{2,2}	4
{2,3}	5
{2,3}	5
{2,5}	7
{2,3}	5
{2,3}	5
{2,5}	7
{3,3}	6
{3, 5}	8
{3,5}	8

The probability distribution for X is given by:

Value of X	$f(x) = P[X = x]$
4	1/10
5	4/10
6	1/10
7	2/10
8	2/10

5.11 (a) Possible values of X are 2, 3, 4, 5, 6, 7, 8, 9, 10, 11, 12.

(b)

$X = 2$	(1,1)					
$X = 3$	(1,2),	(2,1)				
$X = 4$	(1,3),	(2,2),	(3,1)			
$X = 5$	(1,4),	(2,3),	(3,2),	(4,1)		
$X = 6$	(1,5),	(2,4),	(3,3),	(4,2),	(5,1)	
$X = 7$	(1,6),	(2,5),	(3,4),	(4,3),	(5,2),	(6,1)
$X = 8$	(2,6),	(3,5),	(4,4),	(5,3),	(6,2)	
$X = 9$	(3,6),	(4,5),	(5,4),	(6,3)		
$X = 10$	(4,6),	(5,5),	(6,4)			
$X = 11$	(5,6),	(6,5)				
$X = 12$	(6,6)					

(c) The 36 possible outcomes are equally likely so $P[X = x]$ is the number of outcomes for which $X = x$ divided by 36. The probability distribution of X is

x	$f(x)$
2	1/36
3	2/36
4	3/36
5	4/36
6	5/36
7	6/36
8	5/36
9	4/36
10	3/36
11	2/36
12	1/36
Total	1

5.13 (a)

x	$f(x)$
1	0
2	1/6
3	2/6
4	3/6
Total	1

Yes, a probability distribution.

(b)

x	$f(x)$
2	−1/3
3	0
4	1/3
5	2/3
Total	2/3

Not a probability distribution because $f(2)$ is negative, and the values do not sum to 1.

(c)

x	$f(x)$
−2	0
−1	2/20
0	4/20
1	6/20
2	8/20
Total	1

Yes, a probability distribution.

(d)

x	$f(x)$
0	8/15
1	8/30
2	2/15
3	1/15
Total	1

Yes, a probability distribution.

5.15 Since $P(B)=\frac{1}{2}$, $P(C)=\frac{1}{2}$ and purchases in different weeks are independent, the probability model is the same as that for three tosses of a fair coin. Each elementary outcome has probability $\frac{1}{8}$. Therefore

$$P[X=0]=\frac{2}{8},\quad P[X=1]=\frac{4}{8},\quad P[X=2]=\frac{2}{8}.$$

The probability distribution of X is

x	$f(x)$
0	1/4
1	1/2
2	1/4

5.17 Let A be the event "person remembers commercial two hours later." We are given that $P(A) = 0.20$. Take a 4-sample, and let X = number who remember the commercial. Using independence, we derive the probability distribution of X as follows:

x	$f(x)$
0	$P(\bar{A}\bar{A}\bar{A}\bar{A}) = 1(0.80)^4 = 0.4096$
1	$P(A\bar{A}\bar{A}\bar{A}) = 4(0.20)(0.80)^3 = 0.4096$
2	$P(AA\bar{A}\bar{A}) = 6(0.20)^2(0.80)^2 = 0.1536$
3	$P(AAA\bar{A}) = 4(0.20)^3(0.80) = 0.0256$
4	$P(AAAA) = 1(0.20)^4 = 0.0016$
Total	1

5.19 (a) Since $3+1+1+1+1+3=10$, the probabilities of the six faces numbered 1, 2, 3, 4, 5, 6 are $\frac{3}{10}, \frac{1}{10}, \frac{1}{10}, \frac{1}{10}, \frac{1}{10}, \frac{3}{10}$, respectively. The probability distribution of X is

x	$f(x)$
1	0.3
2	0.1
3	0.1
4	0.1
5	0.1
6	0.3
Total	1

(b) $P[\text{Even number}] = f(2) + f(4) + f(6) = 0.1 + 0.1 + 0.3 = 0.5$.

5.21 Since $P[X \text{ is odd}] = f(1) + f(3) + f(5) = 0.1 + 0 + 0.3 = 0.4$ we have $P[X \text{ is even}] = 1 - .4 = .6$, that is, $f(2) + f(4) + f(6) = 0.6$. Now, in order that $f(2)$, $f(4)$ and $f(6)$ are all equal, and their total is 0.6, we must have $f(2) = 0.2$, $f(4) = 0.2$, $f(6) = 0.2$. Thus, the probability distribution of X is given below.

x	$f(x)$
1	0.1
2	0.2
3	0
4	0.2
5	0.3
6	0.2
Total	1

5.23 (a) Consider random selection of one ball from an urn that contains the following mix of 100 numbered balls: 32 balls are numbered 2, 44 balls are numbered 4, and 24 balls are numbered 6. If X denotes the number on the selected ball, then the probability distribution of X would be as given in Table (a).

(b) Consider random selection of one ball from an urn that contains the following mix of 14 numbered balls: 3 balls are numbered -2, 4 balls are numbered 0, 5 balls are numbered 4, and 2 balls are numbered 5. If X denotes the number on the selected ball, then the probability distribution of X would be as given in Table (b).

5.25 (a) $P[X \leq 3] = 0.14 + 0.27 + 0.38 + 0.15 = 0.94$

(b) $P[X \geq 2] = 1 - P[X \leq 1] = 1 - (0.15 + 0.38) = 0.47$

(c) $P[2 \leq X \leq 3] = 0.27 + 0.14 = 0.41$.

5.27 Here $X =$ the number of customers per day.

(a) Customers will be turned away if there are 3 or more customers. The required probability is

$$P[X \geq 3] = f(3) + f(4) + f(5) = 0.25 + 0.15 + 0.05 = 0.45\,.$$

(b) The center's capacity is not fully utilized if fewer than 2 customers arrive. This probability is

$$P[X \leq 1] = f(0) + f(1) = 0.05 + 0.20 = 0.25\,.$$

(c) We see that

$$P[X = 5] = 0.05 \text{ and } P[X \geq 4] = 0.15 + 0.05 = 0.20 > 0.10\,.$$

Therefore, the capacity must be increased by 2. With a capacity of 4, the probability of turning customers away is $P[X = 5] = 0.05$.

5.29 (a) The probability histogram of X is given below.

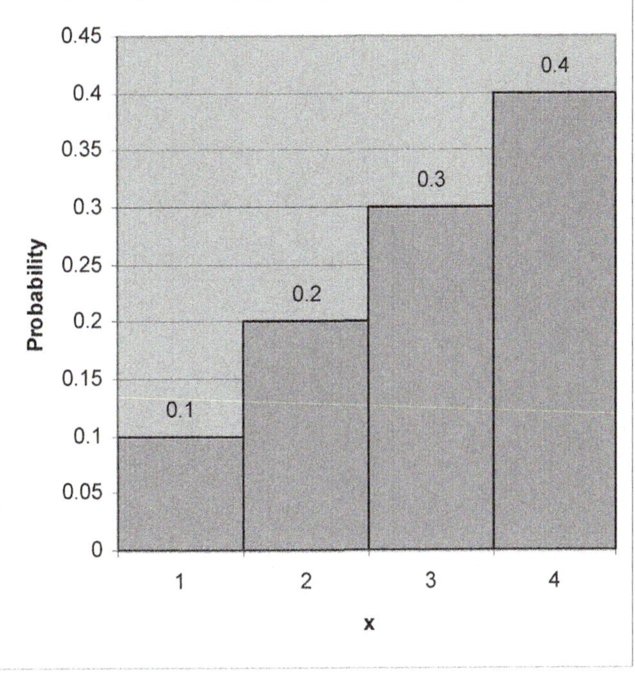

(b) The calculation is given in the following table:

x	$f(x)$	$xf(x)$	$(x-\mu)$	$(x-\mu)^2 f(x)$
1	0.1	0.1	−2	0.4
2	0.2	0.4	−1	0.2
3	0.3	0.9	0	0
4	0.4	1.6	1	0.4
Total		$3.0 = \mu$		$1.0 = \sigma^2$

So, $E(X) = 3, \quad \sigma^2 = 1.0, \quad \sigma = 1.0$

5.31 Let X denote the carpenter's net profit. Then $X = \$5,000$ with probability 0.2 and $X = -\$86$ (loss) with probability $1 - 0.2 = 0.8$. The probability distribution is shown in the table along with the calculation of expectation.

x	$f(x)$	$xf(x)$
−86	0.8	−68.80
5,000	0.2	1000.00
Total	1	$931.20 = E(X)$

So, the expected return $= \$931.20$.

5.33

x	$f(x)$	$xf(x)$	$x^2 f(x)$
0	0.0256	0	0
1	0.1536	0.1536	0.1536
2	0.3456	0.6912	1.3824
3	0.3456	1.0368	3.1104
4	0.1296	0.5184	2.0736
Total		2.4	6.72

$$\mu = 2.4$$

$$\sigma^2 = 6.72 - 2.4^2 = 0.96, \quad \sigma = \sqrt{0.96} = 0.980$$

5.35 (a) We denote "win" by W and "not win" by N, and attach subscripts A or B to identify the project. Listed here are the possible outcomes and calculation of the corresponding probabilities. For instance,

$$P(W_A N_B) = P(W_A)P(N_B), \text{ by independence}$$
$$= 0.50 \times 0.35 = 0.175$$

Outcome	Probability
$W_A W_B$	$0.50 \times 0.65 = 0.325$
$W_A N_B$	$0.50 \times 0.35 = 0.175$
$N_A W_B$	$0.50 \times 0.65 = 0.325$
$N_A N_B$	$0.50 \times 0.35 = 0.175$

(b) & (c) The amounts of profit (X) for the various outcomes are listed below.

Outcome	Profit ($) X
$W_A W_B$	$175,000 + 220,000 = 395,000$
$W_A N_B$	175,000
$N_A W_B$	220,000
$N_A N_B$	0

In the next table, we present the probability distribution of X and calculate $E(X)$.

x	$f(x)$	$xf(x)$	
0	0.175	0	
175,000	0.175	30,625	
220,000	0.325	71,500	
395,000	0.325	128,375	
Total		230,500	$= E(X)$

Expected net profit $= E(X) - \text{cost} = \$230{,}500 - \$2{,}000 = \$228{,}500$.

5.37 (a) & (b) The expectation and standard deviation of X are calculated in the following table.

x	$f(x)$	$xf(x)$	$x^2 f(x)$
0	0.315	0	0
1	0.289	0.289	0.289
2	0.201	0.402	0.804
3	0.114	0.342	1.026
4	0.063	0.252	1.008
5	0.012	0.060	0.300
6	0.006	0.036	0.216
Total		1.381	3.643

$E(X)$ or $\mu = 1.381$

$\text{Var}(X)$ or $\sigma^2 = 3.643 - (1.381)^2 = 1.736$

Standard deviation of X is $\sigma = \sqrt{1.736} = 1.318$.

5.39 (a) To tabulate the probability distribution, we calculate the function $f(x)$ for $x = 0, 1, 2, 3$.

$$f(0) = \frac{1}{84}\binom{5}{0}\binom{4}{3} = \frac{4}{84}$$

$$f(1) = \frac{1}{84}\binom{5}{1}\binom{4}{2} = \frac{5 \times 6}{84} = \frac{30}{84}$$

$$f(2) = \frac{1}{84}\binom{5}{2}\binom{4}{1} = \frac{10 \times 4}{84} = \frac{40}{84}$$

$$f(3) = \frac{1}{84}\binom{5}{3}\binom{4}{0} = \frac{10}{84}$$

x	$f(x)$	$xf(x)$	$x^2 f(x)$
0	4/84	0	0
1	30/84	30/84	30/84
2	40/84	80/84	160/84
3	10/84	30/84	90/84
Total	1	140/84	280/84

(b) Referring to the calculations shown in the table in part (a), we find

$$\text{mean} = \frac{140}{84} = 1.667$$

$$\text{variance} = \frac{280}{84} - \left(\frac{140}{84}\right)^2 = 0.556$$

$$\text{standard deviation } \sigma = \sqrt{0.556} = 0.745.$$

5.41 (a) & (b) Let us use the symbol S for a sale and N for no sale at a customer contact. For contact with 4 customers, we list the elementary outcomes. In assigning the probabilities we use the assumption of independence and the facts that $P(S) = 0.3$, $P(N) = 0.7$. For instance,

$$P(SNNS) = 0.3 \times 0.7 \times 0.7 \times 0.3 = 0.0441.$$

Elementary outcome	Probability	Value of X
$NNNN$	$(0.7)^4 = 0.2401$	0
$NNNS$	$(0.7)^3 \times (0.3) = 0.1029$	1
$NNSN$	0.1029	1
$NSNN$	0.1029	1
$SNNN$	0.1029	1
$NNSS$	$(0.7)^2 \times (0.3)^2 = 0.0441$	2
$NSNS$	0.0441	2
$SNNS$	0.0441	2
$NSSN$	0.0441	2
$SNSN$	0.0441	2
$SSNN$	0.0441	2
$NSSS$	$0.7 \times (0.3)^3 = 0.0189$	3
$SNSS$	0.0189	3
$SSNS$	0.0189	3
$SSSN$	0.0189	3
$SSSS$	$(0.3)^4 = 0.0081$	4
	Total 1.0000	

$P[X=0]=0.2401,$ $\quad P[X=1]=4\times 0.1029=0.4116$

$P[X=2]=6\times 0.0441=0.2646,$ $\quad P[X=3]=4\times 0.0189=0.0756$

$P[X=4]=0.0081$

Probability distribution of X and calculation of expectation

x	$f(x)$	$xf(x)$
0	0.2401	0
1	0.4116	0.4116
2	0.2646	0.5292
3	0.0756	0.2268
4	0.0081	0.0324
Total	1.0000	1.2000

(c) $E(X)=1.2$.

5.43 $P[X\le 2]=0.1+0.2=0.3$ and $P[X\ge 2]=0.2+0.3+0.4=0.9$
The median is 3.

5.45 (a) The Bernoulli model is not appropriate. The assumption of independence is likely to be violated because of peer pressure.

(b) The Bernoulli model is not appropriate because the measurement is on a continuous scale.

(c) The Bernoulli model may be appropriate because there are only two possible outcomes. Independence may be violated for items close together.

(d) Here, each room is a trial with the two possible outcomes: the internet connection was down or it was not. However, the Bernoulli model does not seem plausible because of a lack of independence of the trials.

5.47 Since the event of interest is the drawing of a yellow candy, we identify $S=$yellow, $F=$any color other than yellow.

(a) Because the sampling is with replacement, and each draw has two possible outcomes S or F, the model of Bernoulli trials is appropriate. The mix has 25 candies of which 12 are yellow so the probability of drawing a yellow candy is $\frac{12}{25}=0.48$. We have $p=0.48$.

(b) The model of Bernoulli trials is not appropriate because the sampling is without replacement and the size of the lot is not large. The condition of independence of the trials is violated.

(c) The condition of independence of outcomes in the different trials is violated. For instance, $P(S_2 \mid S_1) = \frac{13}{26}$ while $P(S_2 \mid F_1) = \frac{12}{26}$. The model of Bernoulli trials is not appropriate.

5.49 Label the plots 1, 2, 3 and 4. The number of possible selections of 2 plots out of 4 is $\binom{4}{2} = 6$, and these are equally likely.

(a) Consider the event of an S in the first trial, that is, the first plot is selected. One other plot can be chosen from plots 2, 3, and 4 in $\binom{3}{1} = 3$ ways. Therefore,

$$P(S \text{ in first trial}) = \frac{3}{6} = \frac{1}{2}.$$

The same argument leads to $P(S) = \frac{1}{2}$ in any particular trial.

(b) Denote $S_1S_2 =$ the event that the first and second plots are selected. We have $P(S_1S_2) = \frac{1}{6}$. From part (a), we find that

$$P(S_1)P(S_2) = \frac{1}{2} \times \frac{1}{2} = \frac{1}{4} \neq P(S_1S_2)$$

Therefore, the trials are not independent.

5.51 (a) Although there are two possible outcomes of each trial, the Bernoulli model is not appropriate because the 5 purchases of each consumer cannot be considered independent.

(b) Here the Bernoulli model is plausible because the 500 trials correspond to different consumers who are selected at random.

5.53 We have $P(S) = p = 0.3$, $P(F) = q = 1 - 0.3 = 0.7$.

(a) $P(FFFF) = 0.7^4 = 0.2401$

(b) Because the trials are independent, the required conditional probability is the same as the (unconditional) probability of 4 trials resulting in all successes, which is $P(SSSS) = (0.3)^4 = 0.0081$

(c) $P(FFFS) = (0.7)^3(0.3) = 0.1029$

5.55 (a) $\text{P(SFFSF)} = (0.40)^3(0.60)^2 = 0.02304$

(b) $\text{P(exactly 2 S's)} = \binom{5}{2} \times 0.02304 = 0.2304$

5.57 (a) The possible results in the first two trials are *SS*, *SD*, *DS*, *DD*. If *SS* occurs, the experiment is stopped. With *SD*, there is one more trial so we have either *SDS* or *SDD*. Proceeding in this way, the complete list is

$$\{SS, SDS, SDD, DSS, DSD, DDSS, DDSD, DDDS, DDDD\}$$

(b) We have $P(S) = 1/4,\ P(D) = 3/4$

Outcome	Probability	Value of X
SS	$(1/4)^2 = 1/16$	2
SDS	$(1/4)^2(3/4) = 3/64$	2
SDD	$9/64$	1
DSS	$3/64$	2
DSD	$9/64$	1
DDSS	$9/256$	2
DDSD	$27/256$	1
DDDS	$27/256$	1
DDDD	$81/256$	0

(c) The probability distribution of X is:

x	$f(x)$
0	$81/256$
1	$126/256$
2	$49/256$
Total	1

5.59 (a) Yes. $n = 10,\ p = \frac{1}{6}$

(b) No, because the number of trials is not fixed.

(c) Yes. $n = 3$, and p is the probability of getting a marble numbered either 1 or 2 in a single draw so

$$p = \frac{4+3}{10} = 0.7.$$

(d) No, because X does not represent a count of the number of times that an event occurs.

5.61 (a) With $n=3,\ p=0.35,\ q=0.65$ we obtain

$$P[X=2]=\binom{3}{2}(0.35)^2(0.65)=0.2389$$

(b) $n=6,\ p=0.25,\ q=0.75$

$$P[X=3]=\binom{6}{3}(0.25)^3(0.75)^3=0.132$$

(c) $n=6,\ p=0.65,\ q=0.35$

$$P[X=2]=\binom{6}{2}(0.65)^2(0.35)^4=0.095$$

5.63 $n=4,\ p=0.35$

$$f(x)=\binom{4}{x}(0.35)^x(0.65)^{4-x},\ x=0,\ 1,\ 2,\ 3,\ 4$$

The calculation of $f(x)$ is presented in the following table:

x	$f(x)$	
0	$1(0.65)^4$	$=0.1785$
1	$4(0.35)(0.65)^3$	$=0.3845$
2	$6(0.35)^2(0.65)^2$	$=0.3105$
3	$4(0.35)^3(0.65)$	$=0.1115$
4	$1(0.35)^4$	$=0.0150$
	Total	1.0000

(a) $P[X\le 3]=f(0)+f(1)+f(2)+f(3)=0.985$

(b) $P[X\ge 3]=f(3)+f(4)=0.1265$

(c) $P[X=2 \text{ or } X=4]=f(2)+f(4)=0.3255$

5.65 $n=4,\ p=0.75,\ q=0.25$

$$f(x)=\binom{4}{x}(0.75)^x(0.25)^{4-x},\ \ x=0,1,2,3,4$$

To calculate the required probabilities, it would be convenient to calculate all the $f(x)$ values.

x	$f(x)$	
0	$(0.25)^4$	$= 0.0039$
1	$4(0.75)(0.25)^3$	$= 0.0469$
2	$6(0.75)^2(0.25)^2$	$= 0.2109$
3	$4(0.75)^3(0.25)$	$= 0.4219$
4	$(0.75)^4$	$= 0.3164$
	Total	1.0000

(a) $P[X \geq 3] = f(3) + f(4) = 0.7383$

(b) $P[X \leq 3] = 1 - f(4) = 0.6836$

(c) $E(X) = \sum xf(x) = 0(0.0039) + 1(0.0469) + 2(0.2109) + 3(0.4219) + 4(0.3164) = 3$

5.67 Identify S : severe leaf damage.

X = Number of trees with severe leaf damage in a random sample of 5 trees.

X has a binomial distribution with $n = 5,\ p = 0.15,\ q = 0.85.$

$$f(x) = \binom{5}{x}(0.15)^x(0.85)^{5-x}, \quad x = 0, 1, \ldots, 5$$

(a) $P[X = 3] = f(3) = \binom{5}{3}(0.15)^3(0.85)^2 = 0.024$

(b) $P[X \leq 2] = f(0) + f(1) + f(2) = 0.974.$

5.69 (a) We use the binomial table for $n = 13,\ p = 0.3.$

$$\begin{aligned} P[X = 4] &= P[X \leq 4] - P[X \leq 3] \\ &= 0.654 - 0.421 = 0.233 \end{aligned}$$

(b) '8 failures in 13 trials' means 5 successes in 13 trials. We use the binomial table for $n = 13,\ p = 0.7$

$$P[X = 5] = 0.018 - 0.004 = 0.014$$

(c) Using the binomial table for $n = 13,\ p = 0.3,$ we obtain

$$P[X = 8] = 0.996 - 0.982 = 0.014$$

Refer to (b), and consider interchanging the names 'success' and 'failure'. Then the new p would be .3, (i.e., the old q), and the specified event would then be '8 successes in 13 trials' which is precisely the statement in part (c).

5.71 (a) Denote X = number of successes in 5 trials.

The event 'more than 5 trials are needed in order to obtain 3 successes' means that at most two successes are obtained in 5 trials, that is, $X \leq 2$. Since X has the binomial distribution with $n = 5,\ p = 0.30,$ we use the binomial table to find the required probability

$$P[X \leq 2] = 0.837.$$

(b) The stated event is equivalent to 'at most 4 successes in 9 trials', that is, $X \leq 4$ where X denotes the number of successes in 9 trials. Using the binomial table with $n = 9,\ p = 0.30,$ we find the required probability

$$P[X \leq 4] = 0.901$$

5.73 Mean $= np,\quad sd = \sqrt{npq}$

(a) Mean $= 20 \times 0.75 = 15, \quad sd = \sqrt{20 \times 0.75 \times 0.25} = 1.936$

(b) Mean $= 20 \times 0.10 = 2, \quad sd = \sqrt{20 \times 0.10 \times 0.90} = 1.342$

(c) Mean $= 40 \times 0.30 = 12, \quad sd = \sqrt{40 \times 0.30 \times 0.70} = 2.898$

5.75 Denote X = number of college seniors in support of increased funding, in a random sample of 20 seniors.

Then, X has the binomial distribution with $n = 20,\ p = 0.2.$ We find $P[X \leq 3] = 0.411.$

5.77 $n = 545,\quad p = 0.50,\quad q = 0.50$

mean $= np = 272.5$

$sd = \sqrt{npq} = 11.67$

5.79 (a) X has the binomial distribution with $n = 40$ and

$$p = P[\text{Allergy present}] = 0.16 + 0.09 = 0.25$$

Therefore,

$$E(X) = 40 \times 0.25 = 10$$
$$sd(X) = \sqrt{40 \times 0.25 \times 0.75} = 2.739$$

(b) Y has the binomial distribution with $n = 40$ and

$$p = P[\text{Allergy present} \mid \text{Male}] = \frac{0.16}{0.16 + 0.36} = \frac{16}{52}$$

Therefore,

$$E(Y) = 40 \times \frac{16}{52} = 12.308$$

$$sd(Y) = \sqrt{40 \times \frac{16}{52} \times \frac{36}{52}} = 2.919$$

(c) Z has the binomial distribution with $n = 40$ and

$$p = P[\text{Allergy absent} \mid \text{Female}] = \frac{0.39}{0.09 + 0.39} = \frac{39}{48}$$

Therefore,

$$E(Z) = 40 \times \frac{39}{48} = 32.5$$

$$sd(Z) = \sqrt{40 \times \frac{39}{48} \times \frac{9}{48}} = 2.469$$

5.81 (a) With $n = 20, \quad p = 0.4,$

$$\begin{aligned} P[X \le 4 \text{ or } X \ge 13] &= P[X \le 4] + (1 - P[X \le 12]) \\ &= 0.051 + (1 - 0.979) = 0.072 \end{aligned}$$

(b) $$\begin{aligned} P[X \le 3 \text{ or } X \ge 12] &= P[X \le 3] + (1 - P[X \le 11]) \\ &= 0.016 + (1 - 0.943) = 0.073 \end{aligned}$$

5.83 (a) The center line is at $p_0 = 0.5$, and the lower and upper control limits are

$$0.5 - 3\sqrt{\frac{0.5 \times 0.5}{20}} = 0.165 \text{ and } 0.5 + 3\sqrt{\frac{0.5 \times 0.5}{20}} = 0.835$$

(b) The corresponding proportions 0.4, 0.55, 0.35, 0.75, and 0.5 are graphed below.

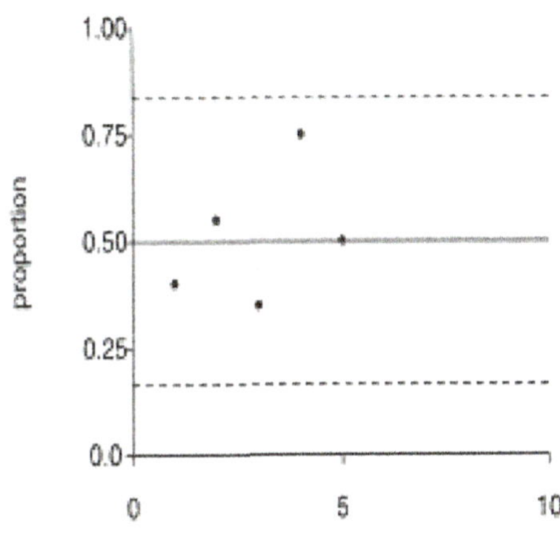

(c) There are no restaurants for which the proportion is out of control.

5.85 Let X be the number of insurance claims handled in one day. Here,

$$f(x) = \frac{e^{-m} m^x}{x!}, \; x = 0,1,2,...$$

(a) $m = 5.5$, so $f(x) = \frac{e^{-5.5} 5.5^x}{x!}$, $x = 0,1,2,...$. Using this, we see that

$$P(X = 6) = f(6) = 0.1571.$$

(b) A rate of 5.5 per day is equivalent to 2(5.5) = 11 in two days. So,

$$f(x) = \frac{e^{-11} 11^x}{x!}, \; x = 0,1,2,...$$

Thus, $P(X = 12) = f(12) = 0.1094$

(c) $m = 5.5$ as in (a). So,

$$\begin{aligned} P(X \le 3) &= f(3) + f(2) + f(1) + f(0) \\ &= 0.1133 + 0.0618 + 0.0225 + 0.0041 \\ &= 0.2017 \end{aligned}$$

5.87 Let X = number of fireflies that flash in a 10 second period. Assume X is Poisson with $m = 1.2$.

(a) P(X = 2) = 0.2169

(b)

$$\begin{aligned} P(X \le 3) &= f(0) + f(1) + f(2) + f(3) \\ &= 0.3012 + 0.3614 + 0.2169 + 0.0867 \\ &= 0.9662 \end{aligned}$$

(c) 1.2 flashings in 10 second period is equivalent to 3(1.2) = 3.6 flashings in a 30-second period.

Let X = number of firefly flashings in 30-second period

Then, X is Poisson with $m = 3.6$. So, $P(X = 6) = 0.0826$.

5.89 Let X = number of alpha particles produced by the decay of radon. X is Poisson with m = 3.

(a) P(X = 5) = 0.1008

(b) Variance = m = 3, standard deviation = $\sqrt{3}$

(c) Three occurrences in one week is equivalent to 3(3) = 9 in three weeks.

Let X = number of alpha particles in three weeks. Then, X is Poisson with m = 9.

$$P(X \le 12) = 0.0001 + 0.0011 + 0.0050 + 0.0150 + 0.0337$$
$$+0.0607 + 0.0911 + 0.1171 + 0.1318 + 0.1318 + 0.1186 +$$
$$0.0970 + 0.0728$$
$$= 0.8758$$

5.91 Let X = number who filed for identity theft in 1 year.
Here, m = 0.0013(500) = 0.65. So, X is Poisson with m = 0.65.

(a) $P(X = 0) = 0.5227$ (b)
$P(X \ge 1) = 1 - P(X = 0) = 1 - 0.5227 = 0.4773$

(c) $P(1 \le X \le 3) = P(X = 1) + P(X = 2) + P(X = 3)$ Observe that

$$P(X = 1) = \frac{0.3293 + 0.3476}{2} = 0.3385$$
$$P(X = 2) = \frac{0.0988 + 0.1217}{2} = 0.1103$$
$$P(X = 3) = \frac{0.0198 + 0.0284}{2} = 0.0241$$

So, $P(1 \le X \le 3) = 0.4729$

5.93 Possible outcomes are: (F,F), (F,M), (M,F), (M,M)
(Here the first entry in each ordered pair refers to the person chosen from List 1, while the second entry refers to the person chosen from List 2.)

Let X = number of females selected in two choices.

(a)

Value of x	Elementary Outcome
0	(M,M)
1	(F,M), (M,F)
2	(F,F)

(b) Now, we compute the probability of each outcome and use this information to produce the probability distribution of X:

$$P(F,F)=\frac{8}{20}\times\frac{10}{30}=\frac{2}{15} \qquad P(M,M)=\frac{12}{20}\times\frac{20}{30}=\frac{2}{5}$$

$$P(F,M)=\frac{8}{20}\times\frac{20}{30}=\frac{4}{15} \qquad P(M,F)=\frac{12}{20}\times\frac{10}{30}=\frac{1}{5}$$

$$P[X=0]=\frac{2}{5}, \qquad P[X=1]=\frac{4}{15}+\frac{1}{5}=\frac{7}{15}, \qquad P[X=2]=\frac{2}{15}$$

Hence, the probability distribution of X is:

x	$f(x)$
0	2/5
1	7/15
2	2/15
Total	1.00

5.95 Let X = number of U.S. companies selected in a 3-sample.
The possible values of X are 0, 1, 2, and 3.

(a) $$P(X=0)=\frac{\binom{3}{0}\binom{7}{3}}{\binom{10}{3}}=\frac{35}{120}=\frac{7}{24} \qquad P(X=1)=\frac{\binom{3}{1}\binom{7}{2}}{\binom{10}{3}}=\frac{3\times 21}{120}=\frac{21}{40}$$

$$P(X=2)=\frac{\binom{3}{2}\binom{7}{1}}{\binom{10}{3}}=\frac{3\times 7}{120}=\frac{7}{40} \qquad P(X=3)=\frac{\binom{3}{3}}{\binom{10}{3}}=\frac{1}{120}$$

The probability distribution of X is

x	$f(x)$
0	7/24
1	21/40
2	21/120
3	1/120
Total	1

(b)

$$\mu=\sum xf(x)=0\left(\frac{7}{24}\right)+1\left(\frac{21}{40}\right)+2\left(\frac{21}{120}\right)+3\left(\frac{1}{120}\right)=0.9$$

$$\sigma^2=\sum x^2 f(x)-\mu^2=\left[0^2\left(\frac{7}{24}\right)+1^2\left(\frac{21}{40}\right)+2^2\left(\frac{21}{120}\right)+3^2\left(\frac{1}{120}\right)\right]-(0.9)^2=0.49$$

$$\sigma=\sqrt{0.49}=0.7$$

5.97 (a)

x	$f(x)$	$xf(x)$
0	0.3	0.0
1	0.4	0.4
2	0.3	0.6
Total	1.0	$1.0 = \mu$

x	$x-\mu$	$(x-\mu)^2$	$f(x)$	$(x-\mu)^2 f(x)$
0	−1	1	0.3	0.3
1	0	0	0.4	0
2	1	1	0.3	0.3
Total				$0.6 = \sigma^2$

$\sigma = \sqrt{0.6} = 0.775$

(b) The probability histogram of X with $\mu = E(X)$ located is

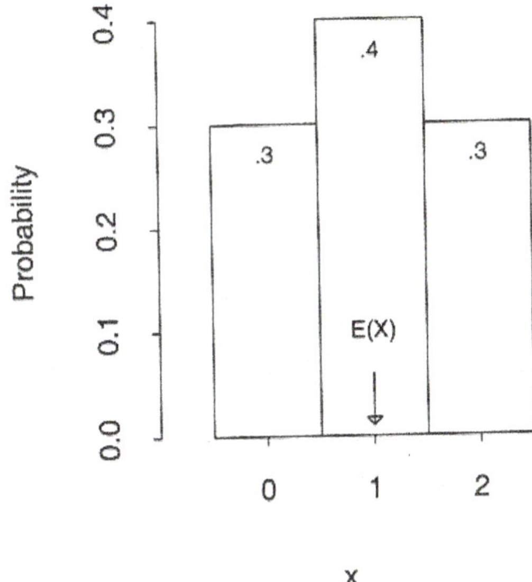

5.99 (a) The probability that the student will get either of the two winning tickets is $\frac{2}{1000} = 0.002$.

(b) Consider $X =$ dollar amount of the student's winnings. The random variable X can have the values 0 or 200 with probabilities 0.998 and 0.002, respectively.

x	$f(x)$	$xf(x)$
0	0.998	0
200	0.002	0.4
		0.4 $= E(X)$

Considering now the purchase price of \$1, the student's expected gain $= \$0.40 - \$1 = -\$0.60$, that is, expected loss $= \$0.60$.

5.101 (a) $P[X < 3] = f(0) + f(1) + f(2) = 0.05 + 0.1 + 0.15 = 0.30.$

(b)

x	$f(x)$	$xf(x)$	$x^2 f(x)$
0	0.05	0	0
1	0.10	0.10	0.10
2	0.15	0.30	0.60
3	0.35	1.05	3.15
4	0.20	0.80	3.20
5	0.15	0.75	3.75
Total		$3.00 = \mu$	10.80

$$E(X) = 3.00$$

$$\sigma^2 = \sum x^2 f(x) - \mu^2 = 10.80 - (3.00)^2 = 1.80$$

$$\sigma = \sqrt{1.80} = 1.342$$

5.103 (a) Let A, B, C denote the correct names. Listed below are the elementary outcomes, that is, the possible assignments of names, and the corresponding values of X = number of matches.

	Correct names			Value of
	A	B	C	X
	A	B	C	3
	A	C	B	1
Possible	B	A	C	1
assignments	B	C	A	0
	C	A	B	0
	C	B	A	1

The probability distribution is presented below along with the calculation of its expectation.

(b) All 6 elementary outcomes are equally likely, Therefore,

$$P[X = 0] = \frac{2}{6}, \quad P[X = 1] = \frac{3}{6}, \quad P[X = 3] = \frac{1}{6}.$$

x	$f(x)$	$xf(x)$
0	2/6	0
1	3/6	3/6
3	1/6	3/6
Total		$1 = E(X)$

5.105 (a) & (b) The possible values of X are:

$X=-15$ if he loses all three times

$X=5-5-5=-5$ if he loses twice and wins once

$X=5+5-5=5$ if he loses once and wins twice

$X=5+5+5=15$ if he wins all three times.

Using the symbol W for win and L for loss at each play we list the elementary outcomes. In assigning probabilities, note that at each play $P(W)=\frac{18}{38}$ and $P(L)=\frac{20}{38}$ because 18 out of 38 slots are red, and 20 are not red. Also the outcomes at different plays are independent.

$X=-15$	$X=-5$	$X=5$	$X=15$
LLL	WLL	WWL	WWW
	LWL	WLW	
	LLW	LWW	

$P(LLL)=\left(\frac{20}{38}\right)^3=0.1458$ so $P[X=-15]=0.1458$

$P(WLL)=\frac{18}{38}\times\frac{20}{38}\times\frac{20}{38}$, and the same result holds for LWL and LLW.

Therefore, $P[X=-5]=3\times\frac{18}{38}\times\left(\frac{20}{38}\right)^2=0.3936$

Similarly, $P[X=5]=3\times\left(\frac{18}{38}\right)^2\times\frac{20}{38}=0.3543$, $P[X=15]=\left(\frac{18}{38}\right)^3=0.1063$.

Probability distribution of X and calculation of expectation

x	$f(x)$	$xf(x)$
−15	0.1458	−2.187
−5	0.3936	−1.968
5	0.3543	1.7715
15	0.1063	1.5945
Total	1.0000	−0.789

$E(X)=-\$0.79$

(c) No. At any play, betting on red or black are probabilistically equivalent because in either case $P(W)=\frac{18}{38}$. Also, different plays are independent.

5.107 (a)

x	$f(x)$	$F(x)$
1	0.07	0.07
2	0.12	0.19
3	0.25	0.44
4	0.30	0.74
5	0.16	0.90
6	0.10	1.00

5.109 (a) Bernoulli model is plausible.

(b) Bernoulli model is plausible.

(c) Not Bernoulli trials because time is continuous.

(d) There are two possible outcomes but the condition of independence does not hold. Clear or cloudy condition often lasts over several days.

(e) Bernoulli model is plausible.

5.111 Denoting male by M and female by F, we calculate

$$P(FFM)=0.5\times 0.5\times 0.5=0.125.$$

5.113 $P(FF)=q^2=\frac{9}{49}$ so $q=\sqrt{\frac{9}{49}}=\frac{3}{7}$ and $p=1-q=\frac{4}{7}$

$$P(SSF)=p^2q=\left(\frac{4}{7}\right)^2\frac{3}{7}=\frac{48}{343}=0.1399$$

5.115 (a) X has the binomial distribution with $n=6$ and $p=0.4$.

(b) Using the binomial table, we find

$$P[X\le 3]=0.821,\ P[X=0]=0.047$$
$$E(X)=np=6\times 0.4=2.4 \text{ persons}$$

5.117 Let X = number of victims under 16 out of 14 moped accident victims. The distribution of X is binomial with $n = 14$ and $p = 0.33$.

(a) Mean $= np = 14 \times 0.33 = 4.62$

(b) $sd = \sqrt{npq} = \sqrt{14 \times 0.33 \times 0.67} = \sqrt{3.095} = 1.759$

(c) P[first victim is under 16 and second victim is at least 16]
$= pq = 0.33 \times 0.67 = 0.221$.

5.119 Employing the binomial model with $n = 20$ and $p = 0.7$, we find $P[X \le 10] =$ 0.048. The probability of the observed result 10, or a more extreme result, is so small that we would doubt the claim that $p = 0.7$. Not as many students as claimed support the paper's view.

5.121 (a) The binomial distribution for $n = 5$ and $p = 0.4$

x	0	1	2	3	4	5	Total
$f(x)$	0.078	0.259	0.346	0.230	0.077	0.010	1.000

(b) The binomial distribution for $n = 5$, $p = 0.4$ is shown below.

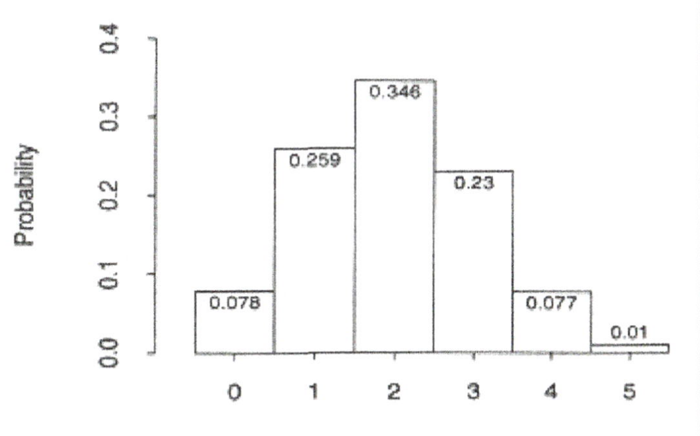

(c) Using part (a) we calculate

x	0	1	2	3	4	5	Total
$xf(x)$	0	0.259	0.692	0.690	0.308	0.050	$1.999 = E(X)$
$x^2 f(x)$	0	0.259	1.384	2.070	1.232	0.250	$5.195 = E(X^2)$

$\text{Var}(X) = 5.195 - (1.999)^2 = 1.198999$

(d) $E(X) = 5 \times 0.4 = 2$

$\text{Var}(X) = 5 \times 0.4 \times 0.6 = 1.2$

The results in part (c) are slightly off due to rounding error.

5.123 Let X = number of persons (out of 20) who feel the system is adequate. Since the city population is large, the binomial distribution is appropriate for X. We have $n = 20$ and $p = 0.3$ so

$$P[X \geq 10] = 1 - P[X \leq 9] = 1 - 0.952 = 0.048$$
$$P[X = 10] = P[X \leq 10] - P[X \leq 9] = 0.983 - 0.952 = 0.031.$$

5.125 Identify S : correct guess

Then $p = P[\text{correct guess in a single trial}] = \frac{1}{5} = 0.2$ because there are 5 possible guesses of which one is correct.

If X denotes the number of correct guesses in 16 trials, the distribution of X is binomial with $n = 16,\ p = 0.2$

(a) Using the binomial table for $n = 16,\ p = 0.2,\ P[X < 8] = 0.999$
(b) "Wrong at least 10 times" means "correct at most $16 - 10 = 6$ times"
The required probability is $P[X \leq 6] = 0.973$.
(c) $E(X) = np = 16 \times 0.2 = 3.2$
$sd(X) = \sqrt{npq} = \sqrt{16 \times 0.2 \times 0.8} = 1.6$.

5.127 Let X = number of bad checks received each day.
X is Poisson with m = 2.

(a) P(X=3) = 0.1804

(b) Variance = m = 2, standard deviation = $\sqrt{2}$

(c) Let Y = number of bad checks received in two consecutive days.
Then, m = 2(2) = 4. So, Y is Poisson with m = 4. Thus, P(Y = 5) = 0.1563.

(d) Let Z = number of bad checks received in three consecutive days.
Then, Z is Poisson with m = 2(3) = 6. Thus,

$$P(Z \leq 10) = f(0) + f(1) + ... + f(10) = 0.9574$$

5.129 (a) The center line is at $p_0 = 0.5$, and the lower and upper control limits are

$$0.5 - 3\sqrt{\frac{0.5 \times 0.5}{20}} = 0.165 \text{ and } 0.5 + 3\sqrt{\frac{0.5 \times 0.5}{20}} = 0.835$$

(b) The corresponding proportions 0.55, 0.4, 0.7, 0.5, 0.65, 0.6, 0.35, 0.7, 0.5 and 0.65 are graphed below.

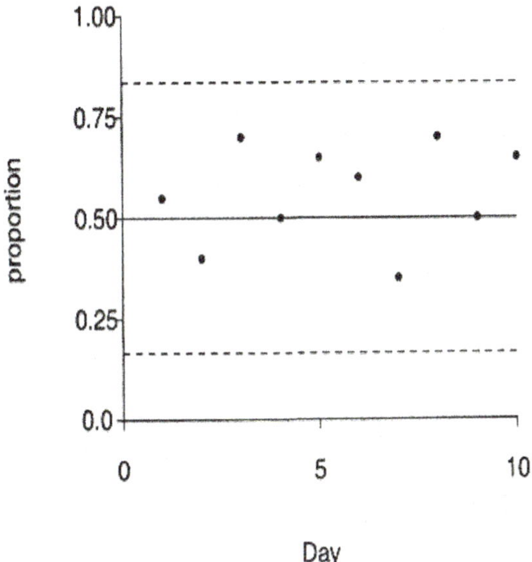

(c) There are no days for which the proportion is out of control.

Chapter 6

THE NORMAL DISTRIBUTION

6.1 (a) The function is non-negative and the area of the rectangle is $0.5 \times 2 = 1$. It is a probability density function.

(b) Since $f(x)$ takes negative values over the interval from 1 to 2, it is not a probability density function.

(c) The function is non-negative and the area of the triangle is $\frac{1}{2} \times base \times height =$ $\frac{1}{2} \times 2 \times 1 = 1$. It is a probability density function.

(d) The function is non-negative, but the area of the rectangle is $1 \times 2 = 2$ so it is not a probability density function.

6.3 From the picture of $f(x)$ we see that the interval 1.5 to 2 has a larger area under $f(x)$ than the interval 0 to 0.5. Consequently, $P[1.5 < X < 2]$ is the larger of the two probabilities.

6.5 For an arbitrary point x in the interval 0 to 2, we find that

$$P[0 < X < x] = \text{Area of the triangle over the base } (0, x) = \frac{1}{2} \cdot x \cdot \frac{x}{2} = \frac{x^2}{4}$$

The median is the value of x for which the cumulative probability $x^2 / 4 = 0.5$ so $x = \sqrt{2} = 1.414$. Similarly, $x = \sqrt{1} = 1$ is Q_1 and $x = \sqrt{3} = 1.73$ is Q_3.

6.7 (a) The median is the time such that there is an equal probability of being earlier or of being later. Consequently, the median is later than 1:20 p.m.

(b) No. The mean could be larger or smaller than the median depending on the distribution. The mean is often larger than the median when the distribution has a long tail to the right.

6.9 The standardized variable is $Z = \dfrac{X-\mu}{\sigma}$. So,

(a) $Z = (X-15)/4$
(b) $Z = (X-61)/9$
(c) $Z = (X-161)/\sqrt{25} = (X-161)/5$

6.11 (a) $Z = (X-25)/3$
(b) $Z = (X-90)/12$

6.13 We use Appendix B Table 4which gives the area under the standard normal curve to the left of a z-value.

(a) $P[Z < 0.63] = 0.7357$
(b) $P[Z < 1.03] = 0.8485$
(c) $P[Z < -1.03] = 0.1515$
(d) $P[Z < -1.35] = 0.0885$

6.15 The area to the right of a z-value $= 1-$ Area to the left of the z-value.

(a) $P[Z > 0.63] = 1 - P[Z < 0.63] = 1 - 0.7357 = 0.2643$
(b) $P[Z > 2.63] = 1 - P[Z < 2.63] = 1 - 0.9957 = 0.0043$
(c) $P[Z > -1.23] = 1 - P[Z < -1.23] = 1 - 0.1093 = 0.8907$
(d) $z = 1.635$ is the mid-point between $z = 1.63$ and $z = 1.64$. From the normal table, we find

$$\begin{array}{ll} P[Z < 1.64] & = 0.9495 \\ P[Z < 1.63] & = 0.9484 \\ \hline \text{difference} & = 0.0011 \end{array}$$

The mid-point between 0.9495 and 0.9484 is

$$0.9484 + \frac{1}{2}(0.0011) = 0.9490 \text{ (rounded)}$$

so $P[Z \le 1.635] = 0.9490$ and $P[Z > 1.635] = 1 - 0.9490 = 0.0510$.

6.17 (a) $P[-0.44 < Z < 0.44] = P[Z < 0.44] - P[Z < -0.44] = 0.6700 - 0.3300 = 0.3400$.
(b) $P[-1.33 < Z < 1.33] = P[Z < 1.33] - P[Z < -1.33] = 0.9082 - 0.0918 = 0.8164$.
(c) $P[0.40 < Z < 2.03] = P[Z < 2.03] - P[Z < 0.40] = 0.9788 - 0.6554 = 0.3234$.
(d) We find that

$$\begin{array}{ll} P[Z<1.41] & = 0.9207 \\ P[Z<1.40] & = 0.9192 \\ \hline \text{difference} & = 0.0015 \end{array}$$

Therefore,

$$P[Z<1.405]=0.9192+\frac{1}{2}(0.0015)=0.9200 \text{ (rounded)}$$

Also,

$$\begin{array}{ll} P[Z<2.31] & = 0.9896 \\ P[Z<2.30] & = 0.9893 \\ \hline \text{difference} & = 0.0003 \end{array}$$

so

$$P[Z<2.306]=0.9893+0.6\times(0.0003)=0.9895 \text{ (rounded)}$$

Finally,

$$P[1.405<Z<2.306]=0.9895-0.9200=0.0695$$

6.19 (a) We are to find the z-value for which the area to the left is 0.20. From the normal table, we find

$$\begin{array}{ll} P[Z<-0.84] & = 0.2005 \\ P[Z<-0.85] & = 0.1977 \\ \hline \text{difference} & = 0.0028 \end{array}$$

Since $0.2005-0.20=0.0005$, the required z-value is

$$Z=-0.84-(0.01)\times\frac{0.0005}{0.0028}=-0.842 \text{ (rounded)}$$

(b) We are to find the z-value for which the area to the left is $1-0.125=0.875$ so $z=1.15$.

(c) By the symmetry of the normal curve, we have

$$2P[Z<-z]=1-P[-z<Z<z]=1-0.668=0.332$$

or

$$P[Z<-z]=0.166, \text{ so that } z=0.97.$$

(d) We are to find the z-value for which

$$0.887=P[z<Z<2.0]=P[Z<2.0]-P[Z<z]=0.9772-P[Z<z]$$

or

$$P[Z<z]=0.0902.$$

We find that

$$P[Z<-1.34]\approx 0.0902$$

So, z = -1.34.

6.21 (a) $P[Z < -0.93] = 0.1762$ so the z-value is -0.93.

(b) We are to find the z-value for which the area to the left is $1 - 0.10 = 0.90$. From the normal table, we find that

$$\begin{array}{ll} P[Z < 1.29] & = 0.9015 \\ P[Z < 1.28] & = 0.8997 \\ \hline \text{difference} & = 0.0018 \end{array}$$

since 0.8997 is nearly 0.90 we could take $z = 1.28$. More accurately, the required z-value is

$$Z = 1.28 + (0.01) \times \frac{0.0003}{0.0018} = 1.28 + 0.0017 = 1.2817 \text{ or } 1.282 \text{ (rounded)}$$

(c) Since $2P[Z < -z] = 1 - 0.954 = 0.046$, we require that

$P[Z < -z] = \frac{1}{2}(0.046) = 0.023$. From the normal table, we see that

$$\begin{array}{ll} P[Z < -1.99] & = 0.0233 \\ P[Z < -2.00] & = 0.0228 \\ \hline \text{difference} & = 0.0005 \end{array}$$

so

$$-1.99 - (0.01) \times \frac{0.0003}{0.0005} = -1.996$$

is the required z-value.

(d) We require that

$$P[Z < z] = 0.50 + P[Z < -0.6] = 0.50 + 0.2743 = 0.7743$$

Scanning the normal table, we find

$$\begin{array}{ll} P[Z < 0.76] & = 0.7764 \\ P[Z < 0.75] & = 0.7734 \\ \hline \text{difference} & = 0.0030 \end{array}$$

The probability 0.7743 is about a third of the way, so we could use 0.753. More accurately, the z-value is

$$Z = 0.75 + (0.01) \times \frac{0.0009}{0.0030} = 0.753$$

6.23 (a) $P[Z < 0.33] = 0.6293$

(b) We find $P[Z < -0.44] = 0.3300$ Therefore, the 33th percentile is -0.44.

(c) $P[Z < 0.97] = 0.8340$

(d) Observe that $P[Z < 1.88] = 0.9699 \approx 0.97$, so that the 97th percentile is approximately 1.88.

6.25 The standardized variable is $Z = \dfrac{X-50}{9}$.

(a) For $x = 46.4$, we have $z = (46.4-50)/9 = -0.4$ so

$$P[X < 46.4] = P[Z < -0.4] = 0.3446$$

(b) $P[X \le 57.2] = P[Z \le 0.8] = 0.7881$

(c) $P[X > 57.2] = P[Z > 0.8] = 1 - P[Z \le 0.8] = 1 - 0.7881 = 0.2119$

(d) $P[X > 60.8] = P[Z > 1.2] = 1 - P[Z \le 1.2] = 1 - 0.8849 = 0.1151$

(e)

$$\begin{aligned} P[33.8 \le X \le 64.4] &= P[\frac{33.8-50}{9} \le Z \le \frac{64.4-50}{9}] \\ &= P[-1.8 \le Z \le 1.6] \\ &= P[Z < 1.6] - P[Z < -1.8] \\ &= 0.9452 - 0.0359 = 0.9093 \end{aligned}$$

(f)

$$\begin{aligned} P[52.5 < X < 60.9] &\approx P[0.28 < Z < 1.21] \\ &= P[Z < 1.21] - P[Z < 0.28] \\ &\approx 0.8869 - 0.6103 = 0.2766 \end{aligned}$$

6.27 The standardized variable is $Z = \dfrac{X-50}{9}$.

(a) From the normal table we find $P[Z < 1.96] = 0.975$. Therefore, $\dfrac{b-50}{9} = 1.96$ so that $b = 50 + 9(1.96) = 67.64$.

(b) Here, $P[X < b] = 1 - 0.025 = 0.975$. As calculated in part (a), we have $b = 67.64$

(c) Since $P[Z < -0.51] = 0.305$, we must have

$$\frac{b-50}{9} = -0.51 \text{ or } b = 50 + 9(-0.51) = 45.41.$$

6.29 Let X denote the score of a randomly selected student. We have $\mu = 500$ and $\sigma = 100$ so the standardized variable is $Z = \dfrac{X-500}{100}$.

(a) $P[X > 650] = P[Z > \dfrac{650-500}{100}] = P[Z > 1.5] = 1 - 0.9332 = 0.0668$

(b) $P[X < 250] = P[Z < -2.5] = 0.0062$

(c) $x = 325$ gives $z = \dfrac{325-500}{100} = -1.75$

$x = 675$ gives $z = \dfrac{675-500}{100} = 1.75$

Thus, we have

$$P[325 < X < 675] = P[-1.75 < Z < 1.75] = 0.9599 - 0.0401 = 0.9198.$$

6.31 The z-value for 32.5 inches is $z = (32.5 - 34.5)/1.4 = -1.429$ and that for 36.5 inches is $z = (36.5 - 34.5)/1.4 = 1.429$ so

$$P[32.5 < X < 36.5] = P[-1.429 < Z < 1.429] = 0.9235 - 0.0765 = 0.847$$

6.33 The z-value for 4 ounces is $(4-5)/1.2 = -0.833$, so that

$$P[X < 4] = P[Z < -0.833] = 0.2024$$

6.35 (a) The z-value for 70 inches is $z = (70 - 64.5)/3.0 = 1.833$. Consequently, by interpolating in the table, we find that

$$P[X > 70] = P[Z > 1.833] = 1 - P[Z < 1.833] = 1 - 0.9664 = 0.0336.$$

(b) The z-value for 60 inches is $z = (60 - 64.5)/3.0 = -1.5$. Consequently (again using interpolation in the table),

$$P[X \le 60] = P[Z \le -1.5] = 0.0668.$$

6.37 The arrival time X is distributed as $N(17, 3)$.

(a) (i) $P[X > 22] = P[Z > \frac{22-17}{3}] = P[Z > 1.67] = 1 - 0.9525 = .0475$

(ii) The z-values corresponding to $x = 13$ and $x = 21$ are $z = \frac{13-17}{3} = -1.33$ and $z = \frac{21-17}{3} = 1.33$, respectively. Thus, we have

$$P[13 < X < 21] = P[-1.33 < Z < 1.33] = 0.9082 - 0.0918 = 0.8164$$

(iii) Since $z = \frac{15.5-17}{3} = -0.5$, $z = \frac{18.5-17}{3} = 0.5$, we have

$$P[15.5 < X < 18.5] = P[-0.5 < Z < 0.5] = 0.6915 - 0.3085 = 0.3830$$

(b) The probability density curve of a normal distribution peaks at the mean. Therefore, the 1-minute interval that has the highest probability is one that is centered at $\mu = 17$ and has length 1, that is, the interval 16.5 to 17.5.

6.39 (a) We use the binomial table (Appendix B Table 2) for $n = 25$ and $p = 0.6$.

(i) $P[X = 17] = P[X \le 17] - P[X \le 16] = 0.846 - 0.726 = 0.120$

(ii) $P[11 \le X \le 18] = P[X \le 18] - P[X \le 10] = 0.926 - 0.034 = 0.892$

(iii) $P[11 < X < 18] = P[X \le 17] - P[X \le 11] = 0.846 - 0.078 = 0.768$

(b) With $n = 25$ and $p = 0.6$, we calculate

$$np = 15 \text{ and } \sqrt{npq} = \sqrt{25 \times 0.6 \times 0.4} = 2.45.$$

We consider X to be normally distributed with mean $= 15$ and $sd = 2.45$, so

$$Z = \frac{X - 15}{2.45} \approx N(0,1)$$

(i) Using the continuity correction, we calculate the normal probability assigned to the interval 16.5 to 17.5.

$$P[16.5 < X < 17.5] = P[\frac{16.5-15}{2.45} < Z < \frac{17.5-15}{2.45}]$$
$$= P[0.612 < Z < 1.020]$$
$$= 0.8461 - 0.7297 = 0.1164.$$

(ii) With continuity correction, the relevant interval is 10.5 to 18.5.

$$P[10.5 < X < 18.5] = P[\frac{10.5-15}{2.45} < Z < \frac{18.5-15}{2.45}]$$
$$= P[-1.837 < Z < 1.429]$$
$$= 0.9235 - 0.0331 = 0.8904.$$

(iii) Here the end points 11 and 18 are not included. The continuity correction leads to the interval 11.5 to 17.5.

$$P[11.5 < X < 17.5] = P[-1.429 < Z < 1.020]$$
$$= 0.8461 - 0.0765 = 0.7696.$$

6.41 Let X = number of adults 18 – 24 who played a musical instrument. Here, p = 0.197 and n = 100. Using normal approximation, we have

$$\frac{X - np}{\sqrt{np(1-p)}} = \frac{X - 19.7}{\sqrt{19.7(0.803)}} = \frac{X - 19.7}{3.977} \approx Z.$$

(a)

$$P[X \le 16] = P[Z \le \frac{16-19.7}{3.977}]$$
$$\approx P[Z \le \frac{16.5-19.7}{3.977}] \text{ (by continuity correction)}$$
$$= P[Z \le -0.8046]$$
$$= 0.5 - 0.2896$$
$$= 0.2104$$

(b) Note: On the first line below, the right hand side should read greater than or equal to 27. This changes the rest of the solution.

$$P[X > 27] = P[X \ge 26]$$
$$= P[Z \ge \frac{26-19.7}{3.977}]$$
$$\approx P[Z > \frac{25.5-19.7}{3.977}] \text{ (by continuity correction)}$$
$$= P[Z > 1.458]$$
$$= 1 - 0.9272 = 0.0728$$
$$= 0.1137$$

(c)

$$\begin{aligned}
P[15 \le X \le 29] &= P[X \le 29] - P[X \le 14] \\
&\approx P[Z \le \frac{29-19.7}{3.977}] - P[Z \le \frac{14-19.7}{3.977}] \\
&\approx P[Z \le \frac{29.5-19.7}{3.977}] - P[Z \le \frac{14.5-19.7}{3.977}] \\
&= P[Z \le 2.464] - P[Z \le -1.308] \\
&= 0.8977
\end{aligned}$$

6.43 (a) Normal approximation is appropriate because n is large and p is not too close to 0 or 1.

(b) Not appropriate because p is too small, $np = 3$.

(c) Not appropriate because p is too close to 1, $n(1-p) = 2.4$.

(d) Normal approximation is appropriate because n is large and p is not too close to 0 or 1.

6.45 The standardized scale is $z = \dfrac{x - np}{\sqrt{np(1-p)}}$. We have the following:

$$\begin{aligned}
n = 5, \quad p = 0.4: \quad & z = (x-2)/1.095 \\
n = 12, \quad p = 0.4: \quad & z = (x-4.8)/1.697 \\
n = 25, \quad p = 0.4: \quad & z = (x-10)/2.449.
\end{aligned}$$

The probability histograms of binomial distribution for $p = 0.4$, $n = 5$, 12, and 25, and the corresponding z-scores are given below.

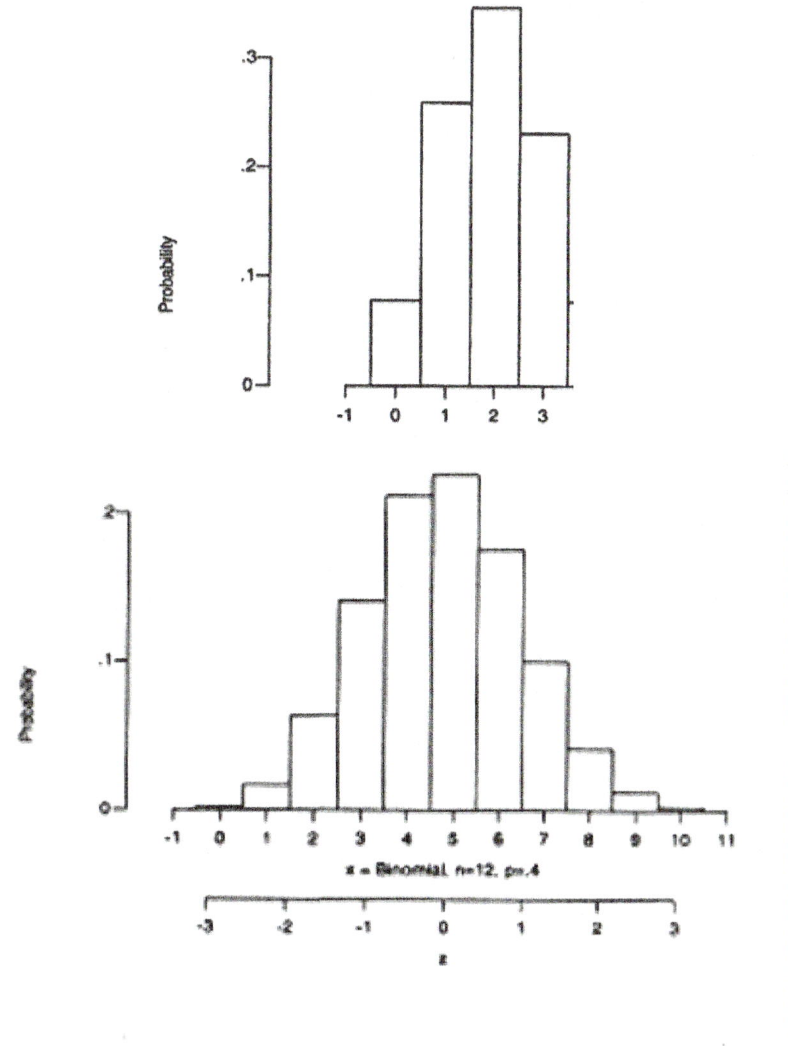
Probability
.3
.2
.1
0
-1 0 1 2 3
Probability
2
.1
0
-1 0 1 2 3 4 5 6 7 8 9 10 11
x = Binomial, n=12, p=.4
-3 -2 -1 0 1 2 3
z

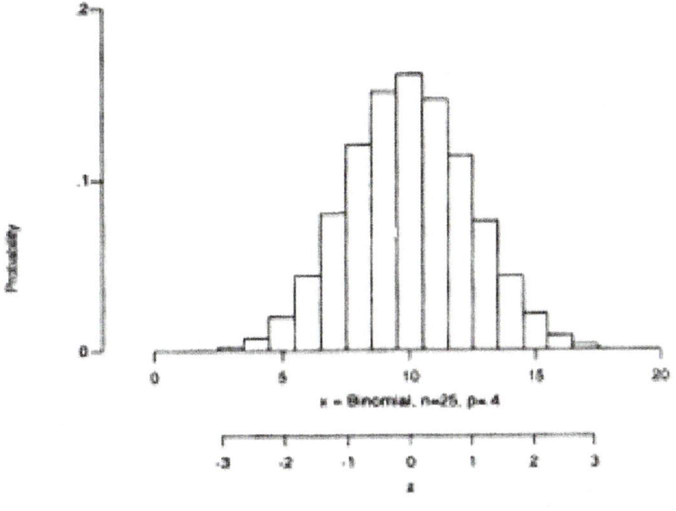
Probability
2
.1
0
0 5 10 15 20
x = Binomial, n=25, p=.4
-3 -2 -1 0 1 2 3
z

6.47 Let X = number of unemployed persons in a random sample of 300. Then, the distribution of X is binomial with $n = 300$ and $p = 0.079$, and a normal approximation is appropriate. Here

$$np = 300 \times 0.079 = 23.7$$

$$\sqrt{npq} = \sqrt{300 \times 0.079 \times 0.921} = 4.672$$

$$Z = \frac{X - 23.7}{4.672}$$

(a) $P[X < 18] \approx P[Z < \frac{17.5 - 23.7}{4.672}] = P[Z < -1.327] = 0.0923$

(b) $P[X > 30] \approx P[Z > \frac{30.5 - 23.7}{4.672}] = P[Z > 1.455] = 1 - 0.9272 = 0.0728$.

6.49 We assume that participants in charity work are independent of one another. The binomial distribution applies. Since $n = 64$ and $p = 0.272$, the distribution of X = number who participate in charity work, is approximately normal with mean $= 64 \times 0.272 = 17.408$ and $sd = \sqrt{64 \times 0.272 \times 0.728} = 3.560$.
Using continuity correction, we approximate

$$P[X \geq 20] = P[Z \geq \frac{19.5 - 17.408}{3.560}] = 1 - P[Z < 0.588] = 1 - 0.7217 = 0.2783$$

6.51 Let Y be next weeks expenditure so

$$p = P[Y > 1500] = P[Z > \frac{1500 - 1450}{220}] = P[Z > 0.2273] = 1 - 0.5899 = 0.4101$$

Let X be the number of weeks, out of $n = 52$, where, the expenses exceed 1500 dollars. Then X has the binomial distribution with $n = 52$ and $p = 0.4101$. To use the normal approximation, we calculate

$\mu = np = 52 \times 0.4101 = 21.3252$ and $\sigma = \sqrt{npq} = \sqrt{52 \times 0.4101 \times 0.5899} = 3.547$.
Using continuity correction,

$$\begin{aligned} P[18 \leq X \leq 24] &\approx P[\frac{17.5 - 21.3252}{3.547} \leq Z \leq \frac{24.5 - 21.3252}{3.547}] \\ &= P[-1.078 < Z < 0.895] \\ &= P[Z < 0.895] - P[Z < -1.078] \\ &= 0.8146 - 0.1418 = 0.6728 \end{aligned}$$

6.53 We assume the residents make their decision to move independently of one another. The binomial distribution applies. Since $n = 100$ and $p = 0.26$, the distribution of X = number who move, is approximately normal with mean $= 100 \times 0.26 = 26$ and $sd = \sqrt{100 \times 0.26 \times 0.74} = 4.386$
Using continuity correction, we approximate

$$P[X \geq 34] \approx P[Z > \frac{33.5 - 26}{4.386}] = 1 - P[Z < 1.71] = 1 - 0.9564 = 0.0436$$

6.55 The large values of volume are too large for the normal distribution to hold. They should be pulled down with a transformation.

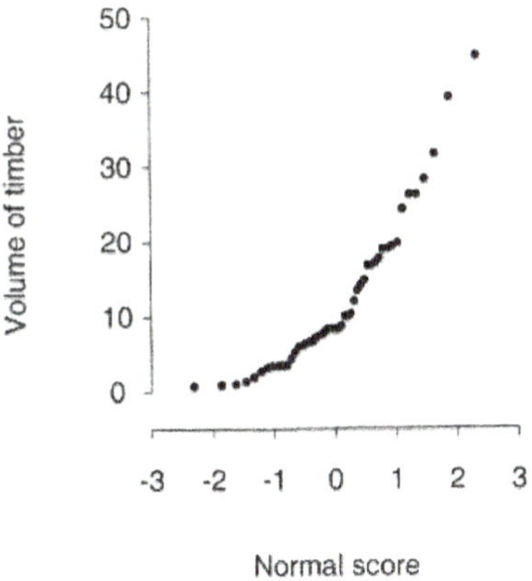

6.57

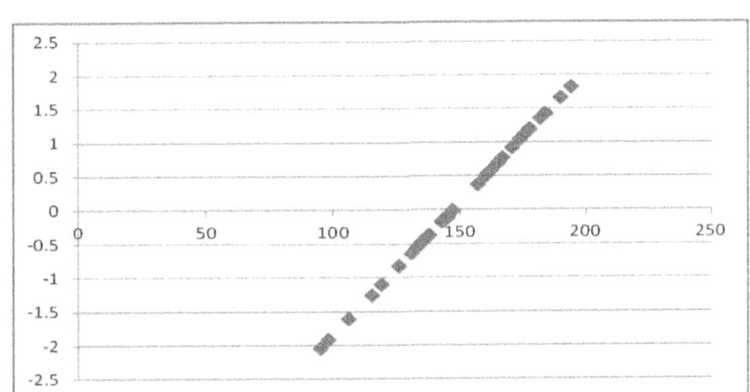

6.59 (a) The normal scores plot for the original data set is as follows:

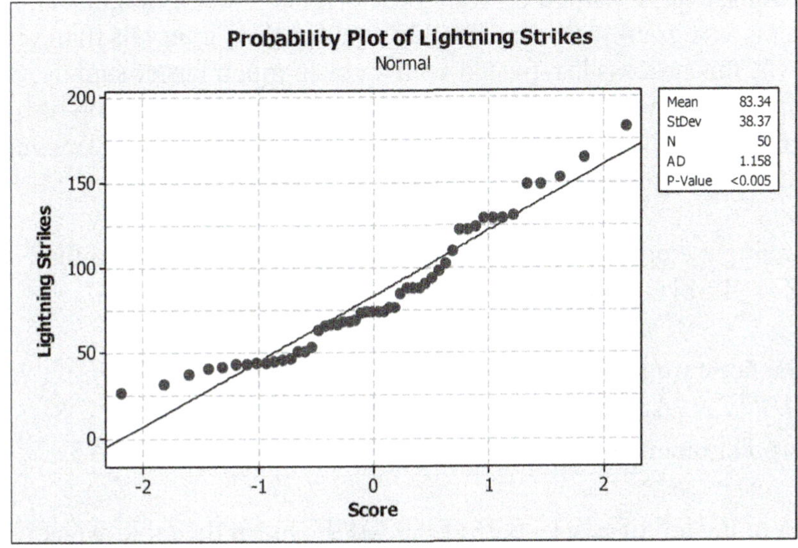

(b) The normal scores plot for the fourth-root of the values in the original data set is as follows:

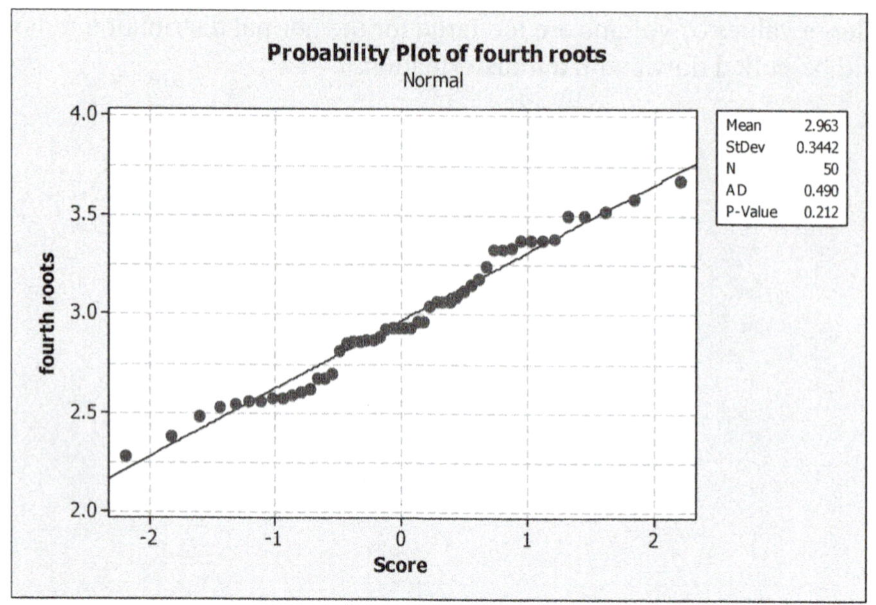

(c) The data in both plots are very close to the line, indicating that a normal approximation in both cases is reasonable.

6.61 The probability density curve is a rectangle with height $= 1$.

(a) The point $x = 0.5$ divides the distribution into halves so median $= 0.5$.

(b) The points $x = 0.25, 0.5$ and 0.75 divide the distribution into quarters. Therefore, first quartile $= 0.25$, second quartile $= 0.5$, and third quartile $= 0.75$.

6.63 We could make a histogram using the first 100 heights and then another based on the first 500 heights. The histogram at the front of the chapter is based on 1,456 observations, a large number. From these original data of heights, one can construct a histogram with considerably smaller class intervals than given in the chapter. If this process is repeated with an even much larger sample, the jumps between consecutive rectangles will then dampen out, and the top of the histogram will approximate the shape of a smooth curve. The density curve would be non-negative and the area under the curve is one.

6.65 (a) Scanning the probabilities given in the normal table, we find that $P[Z < -1.38] = 0.0838$. Therefore, $z = -1.38$.

(b) In the table we look for probabilities close to .047, and find

$$P[Z < -1.68] = 0.0465 \quad \text{and} \quad P[Z < -1.67] = 0.0475$$

Interpolation gives $P[Z < -1.675] = 0.047$, so that $z = -1.675$.

(c) Area to the left of z is $1 - 0.2611 = 0.7389$. From the table we get

$$P[Z < 0.64] = 0.7389\text{, so that } z = 0.64.$$

(d) Area to the left of z is $1-0.12=0.88$. The table gives

$$P[Z<1.17]=0.8790 \text{ and } P[Z<1.18]=0.8810.$$

Since 0.88 is halfway between these results, we interpolate $z=1.175$.

6.67 (a) $P[Z>0.62]=1-P[Z<0.62]=1-0.7324=0.2676$

(b) $P[-1.40<Z<1.40]=P[Z<1.40]-P[Z<-1.40]=0.9192-0.0808=0.8384$

(c) $P\big[|Z|>3\big]=P\big[Z>3\big]+P\big[Z<-3\big]$

$$=2P\big[Z<-3\big] \quad \text{(by symmetry)}$$

$$=2\times 0.0013=0.0026$$

(d) $P\big[|Z|<2\big]=P\big[-2<Z<2\big]=P\big[Z<2\big]-P\big[Z<-2\big]=0.9772-0.0228=0.9544$

Alternatively, we can calculate

$$P\big[|Z|>2\big]=2P\big[Z<-2\big] \quad \text{(as in part (b))}$$

$$=2\times 0.0228=0.0456$$

Hence, $P\big[|Z|<2\big]=1-0.0456=0.9544$.

6.69 The standard normal variable is $Z=\dfrac{X-\mu}{\sigma}=\dfrac{X-280}{58}$.

(a) $P[X<337]<P[\dfrac{337-280}{58}]=P[Z<0.983]=0.8372$

(b) $P[X<240]=P[Z<-0.690]=0.2451$

(c) $P[X>230]=P[Z>-0.862]=1-0.1944=0.8056$

(d) $P[X>90]=P[Z>-3.276]=1-P[Z\le -3.276]=1-0.0005=0.9995$

(e) $P[235<Z<335]=P[-0.776<Z<0.948]=0.8284-0.2188=0.6096$

(f) $P[305<X<405]=P[0.431<Z<2.155]=0.9844-0.6669=0.3175$

6.71 The standardized variable is $Z=\dfrac{X-499}{120}$.

(a) $P[X>600]=P[Z>\dfrac{600-499}{120}]=P[Z>0.842]=1-0.8046=0.1954$

(b) We first find the 90th percentile of the standard normal distribution and then convert it to the x scale. Indeed, observe that

$$P[Z<1.28]=0.8997\approx 0.90$$

The standardized score $z=1.28$ corresponds to

$$x=499+120(1.28)=652.6$$

(c) $P[X<400]=P[Z<-0.83]=0.2033$.

6.73 The strength X is distributed as $N(100,8)$. The bonding will fail if $X<90$. Its probability is

$$P[X < 90] = P[Z < \frac{90-100}{8}] = P[Z < -1.25] = 0.1056.$$

So, about 11% will fail.

6.75 X = number of days of trouble-free operation is normally distributed with $\mu = 750$ and $\sigma = 100$, so the standardized variable is $Z = \frac{X-750}{100}$.

(a) Taking 365 days in a year, 2 years have 730 days.

$$P[X > 730] = P[Z > \frac{730-750}{100}] = P[Z > -0.2]$$
$$= 1 - P[Z < -0.2] = 1 - 0.4207 = 0.5793.$$

(b) Denoting d = number of days of warranty, we require that $P[X < d] = 0.1$. From the standard normal table, we find $P[Z < -1.28] = 0.10$ so

$$\frac{d-750}{100} = -1.28 \text{ or } d = 750 + 100(-1.28) = 622$$

The warranty period can be set to 622 days.

6.77 (a) Denote the volume in one bottle by X which is a random variable having a normal distribution with mean 302 and standard deviation 2 ml. We want to find $P[X < 299]$. Since $Z = \frac{X-302}{2}$ is standard normal, we have

$$P[X < 299] = P[Z < \frac{299-302}{2}] = P[Z < -1.5] = 0.0668$$

(b) We first determine that the z-value 1.645 satisfies $P[Z \le 1.645] = 0.9500$ so $P[Z > 1.645] = 0.05$. Consequently, $v = 302 + (2)1.645 = 305.3$ ounces.

6.79 (a) We use Appendix Table 2 to find the exact binomial probabilities.

(i) When $n = 25$, $p = 0.4$, $P[X \le 7] = 0.154$

(ii) When $n = 20$, $p = 0.7$,

$$P[11 \le X \le 16] = P[X \le 16] - P[X \le 10] = 0.893 - 0.048 = 0.845$$

(iii) When $n = 16$, $p = 0.5$,

$$P[X \ge 9] = 1 - P[X \le 8] = 1 - 0.598 = 0.402.$$

(b) We take X to be approximately normally distributed with mean = np and $sd = \sqrt{npq}$, and use the continuity correction. Also, we round the normal probabilities to three decimals.

(i) $np = 10$, $\sqrt{npq} = \sqrt{25 \times 0.4 \times 0.6} = 2.45$

$$P[X \le 7] = P[Z < \frac{7.5-10}{2.45}] = P[Z < -1.020] = 0.154$$

(ii) $np = 14$, $\sqrt{npq} = 2.05$

$$P[11 \le X \le 16] \approx P[\frac{10.5-14}{2.05} < Z < \frac{16.5-14}{2.05}]$$
$$= P[-1.707 < Z < 1.220] = 0.8891 - 0.0440 = 0.8451$$

(iii) $np = 8$, $\sqrt{npq} = 2$

$$P[X \ge 9] = P[Z > \frac{8.5-8}{2}] = P[Z > 0.25] = 1 - P[Z < 0.25]$$
$$= 1 - 0.5987 = 0.4013$$

6.81 Assuming $n = 400$ is a small fraction of the population, a binomial model is appropriate for X = number of viewers of program A out of 400.

(a) The distribution of X is approximately normal with mean $= np = 400 \times 0.3 = 120$ and $sd = \sqrt{npq} = 9.165$.

$$P[X < 103] \approx P[Z < \frac{102.5-120}{9.165}] = P[Z < -1.91] = 0.0281$$

(b) Under the hypothesis that $p = 0.3$, the calculations in part (a) show that the occurrence of fewer than 103 viewers in a sample of 400 is unlikely (probability 0.0281). Such an observation should strongly support suspicion that $p < 0.3$.

6.83 Out of 400 ticket holders, let X = no. of persons who show up. The distribution of X is binomial with $n = 400$ and $p = 1 - 0.10 = 0.90$. It is approximately normal with mean $= np = 360$ and sd $= \sqrt{npq} = 6$.

(a) One or more reservation holders will not be accommodated if more than 370 passengers show up.

$$P[X > 370] \approx P[Z > \frac{370.5-360}{6}] = P[Z > 1.75] = 1 - 0.9599 = 0.0401$$

(b) $P[X < 350] \approx P[Z < \frac{349.5-360}{6}] = P[Z < -1.75] = 0.0401$.

6.85 Let X be the number of new words in the poem. We approximate the probability that a new word will not be on the list by $14{,}376/884{,}647 = 0.01625$. Then, we treat X as having a binomial distribution with $n = 429$ and $p = 0.01625$.

(a) Expected number of new words $=$ number of words $\times$ probability
$= 429 \times 0.01625 = 6.97$

(b) The standard deviation of X is $\sqrt{429 \times 0.01625 \times 0.98375} = 2.619$. Using the continuity correction, we approximate the binomial probability

$$P[X \geq 12] \approx P[Z \geq \frac{11.5-6.97}{2.619}] = P[Z \geq 1.73] = 1-0.9582 = 0.0418$$

(c) Using continuity correction, we approximate the binomial probability

$$P[X \leq 2] \approx P[Z \leq \frac{2.5-6.97}{2.619}] = P[Z \leq -1.707] = 0.0439$$

(d) $P[2 < X < 12] \approx P[\frac{2.5-6.97}{2.619} \leq Z \leq \frac{11.5-6.97}{2.619}]$

$$= P[Z \leq \frac{11.5-6.97}{2.619}] - P[Z \leq \frac{2.5-6.97}{2.619}]$$

$$= P[Z \leq 1.73] - P[Z \leq -1.707]$$

$$= 0.9582 - 0.0439 = 0.9143$$

The observed number 9 is close to the expected number 6.97. The difference $9-6.97 = 2.03$ is not large according to our probability calculation which shows that ± 4 words from 6.97 is not unusual.

6.87 The normal scores plot is below:

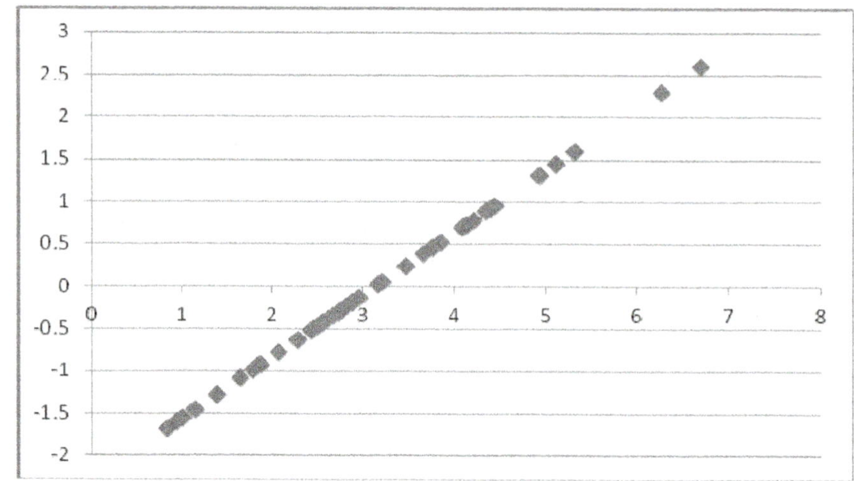

The plot is fairly linear, suggesting the data is approximately normally distributed.

Chapter 7

VARIATION IN REPEATED SAMPLES – SAMPLING DISTRIBUTIONS

7.1 (a) Parameter (b) Statistic (c) Parameter
(d) Statistic (e) Statistic

7.3 (a) Students working full time would be more likely to take an evening course than students who do not work, and so they would be over-represented in the sample. Also, those who work the evening shift are left out of the survey.
(b) A large majority of persons, at all income levels, spend more during the holiday season. The total amount spent and the types of purchases are often atypical. The survey would likely be misleading.

7.5 (a) All possible samples (x_1, x_2) and the corresponding $\bar{x}$ values are:

(x_1, x_2)	(3,3)	(3,5)	(3,7)	(5,3)	(5,5)	(5,7)	(7,3)	(7,5)	(7,7)
$\bar{x} = \frac{x_1+x_2}{2}$	3	4	5	4	5	6	5	6	7

(b) The 9 possible samples are equally likely, so each has a probability 1/9 of occurring. The sampling distribution of $\overline{X}$ is obtained by listing the distinct values of $\overline{X}$ along with the corresponding probabilities, as follows:

$\bar{x}$	Probability $f(\bar{x})$
3	1/9
4	2/9
5	3/9
6	2/9
7	1/9
Total	1

7.7 Not a random sample. The photographer would show his better pictures when trying to get a contract for wedding pictures.

7.9 (a) $X = 1$ if 1 or 2 dots show
$X = 2$ if 3, 4 or 5 dots show
$X = 4$ if 6 dots show

(b) Answers will vary. We obtained the following data from our experiment.

Roll 1 = 3 (so $X = 2$), Roll 2 = 4 (so $X = 2$), Roll 3 = 4 (so $X = 2$)

The median X value is therefore 2.

(c) Answers will vary. The data we collected for the 75 rolls of the die, grouped into 25 3-samples, are tabulated below, along with the corresponding values of X and the median of X in each case.

Rolls	Values of X	Median	Rolls	Values of X	Median
(3,4,4)	(2,2,2)	2	(2,2,2)	(1,1,1)	1
(2,1,6)	(1,1,4)	1	(5,3,2)	(2,2,1)	2
(6,1,1)	(4,1,1)	1	(1,1,2)	(1,1,1)	1
(3,3,4)	(2,2,2)	2	(6,4,6)	(4,2,4)	4
(6,1,6)	(4,1,4)	4	(5,5,2)	(2,2,1)	2
(5,3,2)	(2,2,1)	2	(1,1,6)	(1,1,4)	1
(2,2,6)	(1,1,4)	1	(3,3,5)	(2,2,2)	2
(3,1,3)	(2,1,2)	2	(4,1,2)	(2,1,1)	1
(3,4,1)	(2,2,1)	2	(5,1,1)	(2,1,1)	1
(6,2,2)	(4,1,1)	1	(3,2,3)	(2,1,2)	2
(3,5,5)	(2,2,2)	2	(3,1,5)	(2,1,2)	2
(3,6,4)	(2,4,2)	2	(2,4,4)	(1,2,2)	2
(4,4,3)	(2,2,2)	2			

The relative frequency distribution is:

X	Frequency	Relative Frequency	Population Relative Frequency
1	29	29/75 = 0.38667	0.33333
2	36	36/75 = 0.48000	0.50000
4	10	10/75 = 0.13333	0.16667
Total	75	1.0	1.0

The *sample* relative frequency distribution and the *population* relative frequency distribution should be close since the sample size of 75 is relatively large. In fact, as the sample size increases, the *sample* relative frequency distribution should better approximate the *population* relative frequency distribution.

(d) The sample median values are also listed in the table in part (c). The relative frequency distribution for the median values is as follows:

X_{median}	Frequency	Relative Frequency
1	9	9/25 = 0.36
2	14	14/25 = 0.56
4	2	2/25 = 0.08
Total	25	1.0

This distribution does approximate the sampling distribution since the 25 medians corresponding to the 3-samples of rolls are themselves values of *X*. As such, this is equivalent to rolling a die 25 times, assigning the appropriate value of *X*, and then forming a sampling distribution of *X* based on a sample of size 25. As mentioned above, as the sample size increases, such a *sample* relative frequency distribution should better approximate the *population* relative frequency distribution.

7.11 The solid graph is that of the population density function and the dotted one is for $\overline{X}$ because the sampling distribution will look more normal-like (specifically, more symmetric about the mean) than will the population density.

7.13 We have $\mu = 76$ and $\sigma = 35$.

(a) For $n = 4$, $E(\overline{X}) = \mu = 76$ and $\text{sd}(\overline{X}) = \frac{\sigma}{\sqrt{n}} = \frac{35}{\sqrt{4}} = 17.5$.

(b) For $n = 25$, $E(\overline{X}) = \mu = 76$ and $\text{sd}(\overline{X}) = \frac{\sigma}{\sqrt{n}} = \frac{35}{\sqrt{25}} = 7$.

7.15 The population standard deviation is $\sigma = 1.549$, so $\text{sd}(\overline{X}) = \frac{\sigma}{\sqrt{n}} = \frac{1.549}{\sqrt{n}}$.

(a) For $n = 25$, $\text{sd}(\overline{X}) = \frac{\sigma}{\sqrt{n}} = \frac{1.549}{\sqrt{25}} = 0.3098$

(b) For $n = 100$, $\text{sd}(\overline{X}) = \frac{\sigma}{\sqrt{n}} = \frac{1.549}{\sqrt{100}} = 0.1549$

(c) For $n = 400$, $\text{sd}(\overline{X}) = \frac{\sigma}{\sqrt{n}} = \frac{1.549}{\sqrt{400}} = 0.07745$

(d) In general, $\frac{\sigma}{\sqrt{4n}} = \frac{1}{2}\left(\frac{\sigma}{\sqrt{n}}\right)$. So, quadrupling the sample size cuts $\text{sd}(\overline{X})$ in half.

7.17 We first calculate the mean μ and standard deviation σ of the population that corresponds to *X* taking the values 3, 5, and 7, each having the same probability of occurring (namely 1/3).

x	$f(x)$	$xf(x)$	$x^2 f(x)$
3	1/3	3/3	9/3
5	1/3	5/3	25/3
7	1/3	7/3	49/3
Total	1	15/3	83/3

Using the values in the table, we have the following:

$$\mu = \sum x f(x) = \tfrac{15}{3} = 5$$
$$\sigma^2 = E(X^2) - \mu^2 = \sum x^2 f(x) - \mu^2 = \tfrac{83}{3} - 5^2 = \tfrac{8}{3}, \quad \text{so that } \sigma = \sqrt{\tfrac{8}{3}}$$

For $n = 2$, we know that the mean and standard deviation of the sampling distribution of $\overline{X}$ must be as follows:

$$E(\overline{X}) = \mu = 5$$
$$\text{sd}(\overline{X}) = \tfrac{\sigma}{\sqrt{2}} = \tfrac{\sqrt{\frac{8}{3}}}{\sqrt{2}} = \sqrt{\tfrac{8}{6}} = \sqrt{\tfrac{4}{3}}$$

We verify these by actually calculating the distribution of $\overline{X}$:

$\bar{x}$	$f(\bar{x})$	$\bar{x} f(\bar{x})$	$\bar{x}^2 f(\bar{x})$
3	1/9	3/9	9/9
4	2/9	8/9	32/9
5	3/9	15/9	75/9
6	2/9	12/9	72/9
7	1/9	7/9	49/9
Total	1	45/9	237/9

Using the values in the table, we have the following (which do indeed confirm the above assertion):

$$E(\overline{X}) = \sum \bar{x} f(\bar{x}) = \tfrac{45}{9} = 5$$
$$\text{Var}(\overline{X}) = E(\overline{X}^2) - (E(\overline{X}))^2 = \sum \bar{x}^2 f(\bar{x}) - (E(\overline{X}))^2 = \tfrac{237}{9} - 5^2 = \tfrac{12}{9} = \tfrac{4}{3},$$
$$\text{sd}(\overline{X}) = \sqrt{\tfrac{4}{3}}$$

7.19 (a) All possible samples (x_1, x_2) and the corresponding $\bar{x}$ values and probabilities are tabulated below. By independence, $P(x_1, x_2) = P(x_1) \cdot P(x_2)$, where the values of $P(0)$, $P(1)$, and $P(2)$ are given in the distribution of X. So, for instance, $P(0,0) = P(0) \cdot P(0) = (0.3)(0.3) = 0.09$.

(x_1, x_2)	(0,0)	(0,1)	(0,2)	(1,0)	(1,1)	(1,2)	(2,0)	(2,1)	(2,2)
Probability of (x_1, x_2)	0.09	0.12	0.09	0.12	0.16	0.12	0.09	0.12	0.09
$\bar{x} = \frac{x_1 + x_2}{2}$	0	0.5	1	0.5	1	1.5	1	1.5	2

The sampling distribution of $\overline{X}$ is obtained by listing the distinct values of $\overline{X}$ along with the corresponding probabilities, as follows:

$\overline{x}$	Probability $f(\overline{x})$
0	0.09
0.5	0.24
1	0.34
1.5	0.24
2	0.09
Total	1

(b) $E(\overline{X}) = \mu = \sum x f(x) = 0(0.3) + 1(0.4) + 2(0.3) = 1.0$. This is true for <u>any</u> sample size n.

(c) For $n = 36$, $E(\overline{X}) = 1.0$ (as mentioned in part (b)). Also, $\text{sd}(\overline{X}) = \frac{\sigma}{\sqrt{36}}$, where σ is the *population* standard deviation, which we calculate below.

x	$f(x)$	$xf(x)$	$x^2 f(x)$
0	0.3	0	0
1	0.4	0.4	0.4
2	0.3	0.6	1.2
Total	1	1.0	1.6

$\sigma^2 = E(X^2) - \mu^2 = \sum x^2 f(x) - \mu^2 = 1.6 - 1^2 = 0.6$, so that $\sigma = \sqrt{0.6} = 0.7746$

Thus, $\text{sd}(\overline{X}) = \frac{\sigma}{\sqrt{36}} = \frac{0.7746}{\sqrt{36}} = 0.1291$.

7.21 (a) $E(\overline{X}) = \mu = 115$

(b) $\text{sd}(\overline{X}) = \frac{\sigma}{\sqrt{n}} = \frac{22}{\sqrt{6}} = 8.981$

(c) Since the population distribution is normal, the sample mean $\overline{X}$ has normal distribution with mean 115 and standard deviation 8.981.

7.23 Denote X = weight of a package. We are given that X is normal with mean 32.4 and standard deviation 0.4.

(a) We convert to the standard normal to obtain

$$P[X < 32] = P[\frac{X - 32.4}{0.4} < \frac{32 - 32.4}{0.4}] = P[Z < -1] = 0.1587.$$

Hence, about 16% of the packages weigh less than the labeled amount.

(b) Let X_1 and X_2 denote the weight of two randomly chosen packages. Observe that:

$$E(\overline{X}) = 32.4$$
$$\text{sd}(\overline{X}) = \tfrac{0.4}{\sqrt{2}} = 0.2828$$

Hence, $\overline{X} = \dfrac{X_1 + X_2}{2}$ is normal with mean 32.4 and standard deviation 0.2828.

(c) Again, we convert to the standard normal (using part (b)) to obtain

$$P[\overline{X} < 32] = P[\frac{\overline{X} - 32.4}{0.2828} < \frac{32 - 32.4}{0.2828}] = P[Z < -1.414] = 0.0786.$$

Hence, there is about an 8% chance that the average weight of two packages will be less than the labeled amount of 32 ounces.

7.25 (a) We have $E(\overline{X}) = \mu = 51{,}000$ and $\text{sd}(\overline{X}) = \frac{\sigma}{\sqrt{n}} = \frac{5000}{\sqrt{100}} = 500$. Since $n = 100$ is large, the central limit theorem ensures that the distribution of $\overline{X}$ is approximately normal with mean and standard deviation as calculated above.

(b) The standardized variable is $Z = \dfrac{\overline{X} - 51{,}000}{500}$. As such, we have

$$P[\overline{X} > 51{,}500] = P[Z > \frac{51{,}500 - 51{,}000}{500}] = P[Z > 1] = 0.1587.$$

7.27 The population of fry has mean $\mu = 3.4$ and standard deviation $\sigma = 0.8$, so that

$$E(\overline{X}) = \mu = 3.4 \text{ and } \text{sd}(\overline{X}) = \tfrac{\sigma}{\sqrt{n}} = \tfrac{0.8}{\sqrt{36}} = 0.1333$$

and the standardized variable is $Z = \dfrac{\overline{X} - 3.4}{0.1333}$.

(a) $P[\overline{X} < 3.2] = P[Z < \dfrac{3.2 - 3.4}{0.1333}] = P[Z < -1.5] = 0.0668$

(b) Those caught in the net may be slower, less active fish, or even the less healthy ones. Consequently, they may tend to be on the smaller side of the distribution.

7.29 We have $E(\overline{X}) = \mu = 34.5$ and $\text{sd}(\overline{X}) = \frac{\sigma}{\sqrt{n}} = \frac{1.3}{\sqrt{6}} = 0.5307$, and the standardized variable is $Z = \dfrac{\overline{X} - 34.5}{0.5307}$. As such, we have

$$\begin{aligned} P[34.1 < \overline{X} < 35.2] &= P[\frac{34.1 - 34.5}{0.5307} < Z < \frac{35.2 - 34.5}{0.5307}] \\ &= P[-.7537 < Z < 1.319] \\ &= P[Z < 1.319] - P[Z < -0.7537] = 0.9064 - 0.2255 = 0.6809 \end{aligned}$$

7.31 (a) By column, the medians are as follows:

6	4	4	9	6	6	6
7	4	6	4	4	1	6
4	4	4	5	7	6	5
6	4	7	4	6	7	1
4	6	4	2	5	9	5

4	2	4	7	5	3	8
4	2	2	2	4	7	2
7	5	5	4	5	2	4
6	6	5	4	4	4	3
6	2	4	5	4	4	2

5	3	6	4	1	7
3	5	4	4	5	3
5	8	4	6	6	4
6	4	4	8	3	3
5	5	2	7	5	6

(b) & (c) Histograms are given below. The mean has smaller variance.

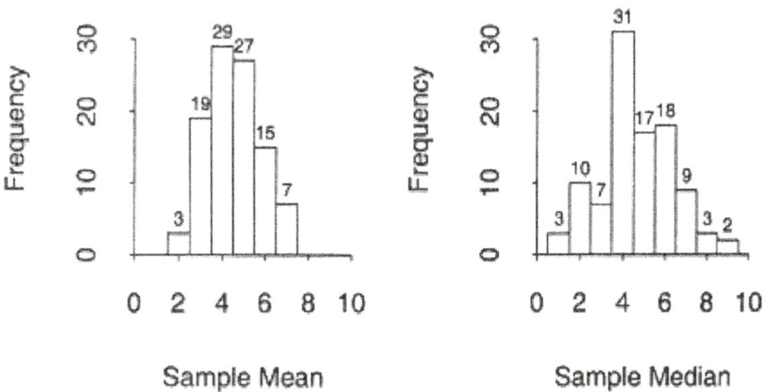

Frequency Chart:

x	1	2	3	4	5	6	7	8	9
frequency	3	10	7	31	17	18	9	3	2

7.33 (a) and (b) We use the following tabulated values to calculate the mean and standard deviation:

x	$f(x)$	$x f(x)$	$x^2 f(x)$
0	0.4	0	0
1	0.3	0.3	0.3
2	0.1	0.2	0.4
3	0.2	0.6	1.8
Total	1	1.1	2.5

Consequently, $E(X) = 1.1$ and $\text{sd}(X) = \sqrt{2.5 - (1.1)^2} = \sqrt{1.29} = 1.136$.

(c) Let X_1 be the number of complaints on the first day and X_2 the number of complaints on the second day. The total number of complaints is the sum $X_1 + X_2$. In order to determine the probability distribution of the sum, we make use of the fact that X_1 and X_2 are independent. For instance, the outcome (1,2), the total is $1 + 2 = 3$ and, by independence, the probability of the outcome (1,2) is the product $(0.3)(0.1) = 0.03$. All possible samples (x_1, x_2), along with the corresponding totals and probabilities, are tabulated below:

(x_1, x_2)	(0,0)	(0,1)	(0,2)	(0,3)	(1,0)	(1,1)	(1,2)	(1,3)
Probability of (x_1, x_2)	0.16	0.12	0.04	0.08	0.12	0.09	0.03	0.06
$x_1 + x_2$	0	1	2	3	1	2	3	4

(x_1, x_2)	(2,0)	(2,1)	(2,2)	(2,3)	(3,0)	(3,1)	(3,2)	(3,3)
Probability of (x_1, x_2)	0.04	0.03	0.01	0.02	0.08	0.06	0.02	0.04
$x_1 + x_2$	2	3	4	5	3	4	5	6

The sampling distribution of $X_1 + X_2$ is obtained by listing the distinct values of $X_1 + X_2$ along with the corresponding probabilities, as follows:

$x_1 + x_2$	Probability
0	0.16
1	0.24
2	0.17
3	0.22
4	0.13
5	0.04
6	0.04
Total	1

(d) Note that if the number of complaints is more than 125, the sample mean will be greater than $\frac{125}{90}$. Since the sample size $n = 90$ is large, we approximate the distribution of $\overline{X}$ by a normal distribution with mean 1.1 and standard deviation $\frac{\sigma}{\sqrt{n}} = \frac{1.136}{\sqrt{90}} = 0.1197$. Indeed, we have

$$P[\overline{X} > \tfrac{125}{90}] = P[Z > \frac{\frac{125}{90} - 1.1}{0.1197}] = P[Z > 2.413] = 0.0079.$$

7.35

(a) The histogram is given below:

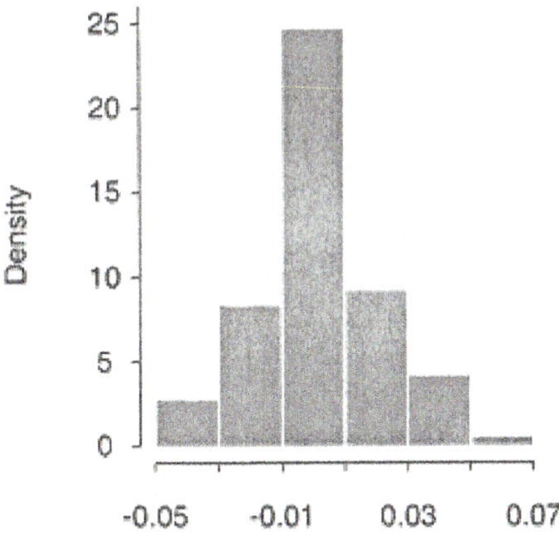

(b) There is more variability in the differences two months apart than in the one-month differences. A computer calculation gives the standard deviation 0.0.0275 for two-month differences versus 0.0234 for the one-month differences.

7.37 Yes, the solid one could be the population density and the dotted one could be that of the graph of the sampling distribution (for small n).

7.39 (a) By column, we record here the value of R for each possible sample listed in Exercise 7.34(a).

0	2	4	6	2	0	2	4
4	2	0	2	6	4	2	0

(b) For the sampling distribution of R, we list the distinct values along with the corresponding probabilities:

Value of R	Probability
0	4/16
2	6/16
4	4/16
6	2/16
Total	1

7.41 (a) All possible samples (x_1, x_2) and the corresponding $\bar{x}$ values and probabilities are tabulated below. By independence, $P(x_1, x_2) = P(x_1) \cdot P(x_2)$, where the values of $P(1)$, $P(2)$, and $P(3)$ are given in the distribution of X. So, for instance, $P(1,1) = P(1) \cdot P(1) = (0.2)(0.2) = 0.04$.

(x_1, x_2)	(1,1)	(1,2)	(1,3)	(2,1)	(2,2)	(2,3)	(3,1)	(3,2)	(3,3)
Probability of (x_1, x_2)	0.04	0.12	0.04	0.12	0.36	0.12	0.04	0.12	0.04
$\bar{x} = \frac{x_1 + x_2}{2}$	1	1.5	2	1.5	2	2.5	2	2.5	3

The sampling distribution of $\overline{X}$ is obtained by listing the distinct values of $\overline{X}$ along with the corresponding probabilities, as follows:

$\bar{x}$	Probability $f(\bar{x})$
1	0.04
1.5	0.24
2	0.44
2.5	0.24
3	0.04
Total	1

(b) $E(\overline{X}) = \mu = \sum x f(x) = 1(0.2) + 2(0.6) + 3(0.2) = 2.0$. This is true for <u>any</u> sample size n.

(c) For $n = 81$, $E(\overline{X}) = 2.0$ (as mentioned in part (b)). Also, $\text{sd}(\overline{X}) = \frac{\sigma}{\sqrt{81}}$, where σ is the *population* standard deviation, which we calculate below.

x	$f(x)$	$xf(x)$	$x^2 f(x)$
1	0.2	0.2	0.2
2	0.6	1.2	2.4
3	0.2	0.6	1.8
Total	1	2.0	4.4

$\sigma^2 = E(X^2) - \mu^2 = \sum x^2 f(x) - \mu^2 = 4.4 - 2^2 = 0.4$, so that $\sigma = \sqrt{0.4} = 0.6325$

Thus, $\text{sd}(\overline{X}) = \frac{\sigma}{\sqrt{81}} = \frac{0.6325}{\sqrt{81}} = 0.0703$.

7.43 We have $\mu = 32.4$, $\sigma = 0.4$, and $n = 9$.

(a) $E(\overline{X}) = \mu = 32.4$, $\text{sd}(\overline{X}) = \frac{\sigma}{\sqrt{n}} = \frac{0.4}{\sqrt{9}} = 0.1333$

(b) Since the population distribution is normal, the sample mean $\overline{X}$ has normal distribution with mean and standard deviation given in part (a).

(c) The standardized variable is $Z = \dfrac{\overline{X} - 32.4}{0.1333}$. As such, we have

$$P[32.3 < \overline{X} < 32.6] = P[\frac{32.3 - 32.4}{0.1333} < Z < \frac{32.6 - 32.4}{0.1333}] = P[-0.750 < Z < 1.500]$$
$$= P[Z < 1.500] - P[Z < -0.750] = 0.9332 - 0.2266 = 0.7066.$$

7.45 We have $\mu = 12.1$, $\sigma = 3.2$, and $n = 9$.

(a) Since the population is normal, the distribution of $\overline{X}$ is normal with mean $= \mu = 12.1$ and sd $= \frac{\sigma}{\sqrt{n}} = \frac{3.2}{\sqrt{9}} = \frac{3.2}{3}$.

(b) The standardized normal variable is $Z = \dfrac{\overline{X} - 12.1}{3.2/3}$. As such, observe that

$$P[\overline{X} < 10] = P[Z < \frac{10 - 12.1}{3.2/3}] = P[Z < -1.97] = 0.0244.$$

(c) The pebble size X is normally distributed with mean 12.1 and standard deviation 3.2, so $Z = \dfrac{X - 12.1}{3.2}$ is $N(0,1)$. As such, observe that

$$P[X < 10] = P[Z < \frac{10 - 12.1}{3.2}] = P[Z < -0.656] = 0.256.$$

So, about 26% of the pebbles are of size smaller than 10.

7.47 We have $\mu = 1.9$, $\sigma = 1.2$, and $n = 36$.

(a) $E(\overline{X}) = \mu = 1.9$ and $\text{sd}(\overline{X}) = \frac{\sigma}{\sqrt{n}} = \frac{1.2}{\sqrt{36}} = 0.2$.

(b) Since $n = 36$ is large, the central limit theorem ensures that the distribution of $\overline{X}$ is approximately normal with mean and standard deviation as calculated in part (a).

7.49 (a) We need to first compute μ and σ:

x	$f(x)$	$xf(x)$	$x^2 f(x)$
1	0.02	0.02	0.02
2	0.02	0.04	0.08
3	0.04	0.12	0.36
4	0.12	0.48	1.92
5	0.80	4.0	20
Total		4.66	22.38

Thus,

$$\mu = \sum xf(x) = 4.66$$
$$\sigma^2 = \sum x^2 f(x) - \mu^2 = 22.38 - 4.66^2 = 0.6644$$
$$\sigma = \sqrt{0.6644} = 0.815$$

Hence, $E\left(\overline{X}\right) = \mu = 4.66$ and $sd\left(\overline{X}\right) = \frac{\sigma}{\sqrt{40}} = \frac{0.815}{\sqrt{40}} = 0.1289$.

(b) Since n is large, $\overline{X}$ has approximately a normal distribution with mean and s.d. given in part (a).

(c)

$$P(4.6 \le \overline{X} \le 4.8) = P(\frac{4.6-4.66}{0.1289} \le Z \le \frac{4.8-4.66}{0.1289})$$
$$= P(-0.4655 \le Z \le 1.086)$$
$$= P(Z \le 1.086) - P(Z \le -0.4655)$$
$$= 0.8660 - 0.3660 = 0.5$$

7.51 (a) & (b) We calculate the population mean and standard deviation using the tabulated values below:

x	$f(x)$	$xf(x)$	$x^2 f(x)$
0	0.5	0	0
1	0.3	0.3	0.3
2	0.2	0.4	0.8
Total	1	0.7	1.1

Using the values in the table, we have the following:

$\mu = \sum xf(x) = 0.7$

$\sigma^2 = E(X^2) - \mu^2 = \sum x^2 f(x) - \mu^2 = 1.1 - (0.7)^2 = 0.61$, so that $\sigma = \sqrt{0.61} = 0.781$

(c) Let X_1 be the number sold the next day and X_2 the number sold the day after next. The total number sold is the sum $X_1 + X_2$. In order to determine the probability distribution of the sum, we make use of the fact that X_1 and X_2 are independent. For instance, the outcome (2,1), the total is $2+1=3$ and, by independence, the probability of the outcome (2,1) is the product $(0.2)(0.3) = 0.06$. All possible samples (x_1, x_2), along with the corresponding totals and probabilities, are tabulated below:

(x_1, x_2)	(0,0)	(0,1)	(0,2)	(1,0)	(1,1)	(1,2)	(2,0)	(2,1)	(2,2)
Probability of (x_1, x_2)	0.25	0.15	0.10	0.15	0.09	0.06	0.10	0.06	0.04
$x_1 + x_2$	0	1	2	1	2	3	2	3	4

The sampling distribution of $X_1 + X_2$ is obtained by listing the distinct values of $X_1 + X_2$ along with the corresponding probabilities, as follows:

$x_1 + x_2$	Probability
0	0.25
1	0.30
2	0.29
3	0.12
4	0.04
Total	1

(d) The event that at least 53 kayaks are sold is the same event that $\overline{X} > \frac{53}{64}$. Since the sample size $n = 64$ is large, we approximate the distribution of $\overline{X}$ by a normal distribution with mean 0.7 and standard deviation $\frac{\sigma}{\sqrt{n}} = \frac{0.781}{\sqrt{64}} = 0.09763$.

Indeed, we have

$$P[\overline{X} > \tfrac{53}{64}] = P[Z > \frac{\frac{53}{64} - 0.7}{0.09763}] = P[Z > 1.312] = 0.0948.$$

(e) Since $P[Z > 1.645] = 0.0500$, we must have the z-value 1.645. If k is the required number, then

$$Z = \frac{\frac{k}{64} - 0.7}{0.09763} = 1.645,$$

so that $k = 64(0.7 + 1.645(0.09763)) = 55.1$. Hence, 56 kayaks must be ordered.

7.53 (a) Let X be the amount in a single bottle. The distribution of X is normal, so that the desired probability is

$$P[X < 299] = P[\frac{X - 302}{2} < \frac{299 - 302}{2}] = P[Z < -1.5] = 0.0668.$$

(b) Since $P[Z > 1.645] = 0.0500$, we have

$$Z = \frac{v - 302}{2} = 1.645 \quad \text{or} \quad v = 302 + 2(1.645) = 305.29 \text{ ml}.$$

(c) Since the population is normal, $\overline{X}$ has a normal distribution with mean 302 and standard deviation $\frac{\sigma}{\sqrt{n}} = \frac{2}{\sqrt{2}} = \sqrt{2}$. Indeed, we have

$$P[\overline{X} < 299] = P[Z < \frac{299 - 302}{\sqrt{2}}] = P[Z < -2.121] = 0.017.$$

(d) From part (c) each package has probability of 0.017 of containing less than 299 ml. Let Y be the number out of 2 packages that contain less than 299 ml. Then, Y has a binomial distribution with $n = 2$ and $p = 0.017$ We are interested in the event $Y = 1$. By the formula for the binomial distribution, we see that the probability of Y occurring is:

$$2(0.017)(1 - 0.017) = 0.0334.$$

Chapter 8

DRAWING INFERENCES FROM LARGE SAMPLES

8.1 (a) (i) $S.E. = \frac{\sigma}{\sqrt{n}} = \frac{65}{\sqrt{125}} = 5.814$

(ii) 98% error margin is $z_{0.01}\frac{\sigma}{\sqrt{n}} = (2.33)(5.814) = 13.547$.

(b) (i) $S.E. = \frac{\sigma}{\sqrt{n}} = \frac{38}{\sqrt{47}} = 5.543$

(ii) 95% error margin is $z_{0.025}\frac{\sigma}{\sqrt{n}} = (1.96)(5.543) = 10.864$.

(c) (i) $S.E. = \frac{\sigma}{\sqrt{n}} = \frac{20}{\sqrt{6}} = 8.165$

(ii) 90% error margin is $z_{0.05}\frac{\sigma}{\sqrt{n}} = (1.645)(8.165) = 13.431$.

8.3 (a) point estimate of μ is $\overline{x} = 63.6$ pounds.

Estimated standard error is $\frac{s}{\sqrt{n}}$, where $s^2 = \frac{\sum(x_i - \overline{x})^2}{n-1}$, is approximately 66.7.

(b) point estimate of μ is $\overline{x} = 79.4$ pounds.

Estimated standard error is $\frac{s}{\sqrt{n}}$, where $s^2 = \frac{\sum(x_i - \overline{x})^2}{n-1}$, is approximately 45.49.

8.5 Estimated mean weakly amount contested is $\overline{x} = \$95.74$.

Estimated standard error is $\frac{s}{\sqrt{n}} = \frac{24.63}{\sqrt{50}} = \$3.483$.

90% error margin is $z_{0.05}\frac{s}{\sqrt{n}} = 1.645(3.483) = \$5.730$.

8.7 (a) Given that the 95% error margin is $1.96\frac{s}{\sqrt{n}} = 4.2$, we know that the estimated

standard error is $\frac{s}{\sqrt{n}} = \frac{4.2}{1.96} = 2.143$.

(b) 90% error margin is $1.645\frac{s}{\sqrt{n}} = 1.645(2.143) = 3.525$.

8.9 Recall that the 95% error margin is $1.96\frac{\sigma}{\sqrt{n}}$.

(a) In order that the 95% error margin be equal to $\frac{1}{8}\sigma$, we must have

$$1.96\frac{\sigma}{\sqrt{n}} = \tfrac{1}{8}\sigma,$$

so that solving for n yields $n = (1.96 \times 8)^2 = 245.9$. So, the required sample size is $n = 246$.

(b) In order that the 95% error margin be equal to 0.15σ, we must have

$$1.96\frac{\sigma}{\sqrt{n}} = 0.15\sigma,$$

so that solving for n yields $n = \left(\frac{1.96}{0.15}\right)^2 = 170.7$. So, the required sample size is $n = 171$.

8.11 We have $d = 5.00$, $\sigma = 25$, and $z_{\alpha/2} = z_{0.01} = 2.326$. Hence,

$$\left[\frac{2.326(25)}{5.00}\right]^2 = 135.26.$$

So, the required sample size is $n = 135$.

8.13 We have $d = 0.5$, $\sigma = 2.5$, and $z_{\alpha/2} = z_{0.05} = 1.645$. Hence,

$$\left[\frac{1.645(2.5)}{0.5}\right]^2 = 67.6.$$

So, the required sample size is $n = 68$.

8.15 Observe that $1 - \alpha = 0.98$ implies $\alpha = 0.02$, so that $z_{\alpha/2} = z_{0.01} = 2.326$. A 98% confidence interval for μ is calculated as follows:

$$\bar{x} \pm z_{0.01}\frac{s}{\sqrt{n}} = 2.583 \pm 2.326\frac{1.303}{\sqrt{30}} = 2.583 \pm 0.5533 \text{ or } (2.0297, 3.136).$$

8.17 95% of 365, namely 347, confidence intervals are expected to cover the true means. Before the data are obtained each day, the probability of covering the population mean is 0.95. By the long run frequency interpretation of probability, approximately 95% will cover, assuming each day of the year is a representative sample.

8.19 For large n, a 95% confidence interval for μ is given by $\overline{X} \pm z_{0.025}\frac{S}{\sqrt{n}}$. Using $z_{0.025} = 1.96$, $n = 35$, and the summary statistics $\overline{x} = 30.2$ grams, $s = 3.8$ grams, the 95% confidence interval for μ is given by

$$30.2 \pm 1.96\frac{3.8}{\sqrt{35}} = 30.2 \pm 1.26 \text{ or } (28.94, 31.46) \text{ grams.}$$

8.21 For large n, a 99% confidence interval for μ is given by $\overline{X} \pm z_{0.005}\frac{S}{\sqrt{n}}$. Using $z_{0.005} = 2.58$, $n = 120$, and the summary statistics $\overline{x} = 18.3$ days, $s = 5.2$ days, the 99% confidence interval for μ (true mean survival time) is given by

$$18.3 \pm 2.58\frac{5.2}{\sqrt{120}} = 18.3 \pm 1.2 \text{ or } (17.1, 19.5) \text{ days.}$$

8.23 (a) For large n, a 95% confidence interval for μ is given by $\overline{X} \pm z_{0.025}\frac{S}{\sqrt{n}}$. Using $z_{0.025} = 1.96$, $n = 49$, and the summary statistics $\overline{x} = 3.8$, $s = 0.7$, the 95% confidence interval for μ is given by

$$3.8 \pm 1.96\frac{0.7}{\sqrt{49}} = 3.8 \pm 0.196 \text{ or } (3.604, 3.996).$$

(b) This is uncertain. We only know that in the long run, 95% of all confidence intervals would contain the value of the true parameter μ.

(c) 95% of all confidence intervals would contain μ by the interpretation provided in the text.

8.25 For large n, a 90% confidence interval for μ is given by $\overline{X} \pm z_{0.05}\frac{S}{\sqrt{n}}$. Using $z_{0.05} = 1.645$, $n = 140$, and the summary statistics $\overline{x} = 8.6$ miles, $s = 4.3$ miles, the 90% confidence interval for μ (true mean commuting distance) is given by

$$8.6 \pm 1.645\frac{4.3}{\sqrt{140}} = 8.6 \pm 0.60 \text{ or } (8.00, 9.20) \text{ miles.}$$

8.27 For large n, a 95% confidence interval for μ is given by $\overline{X} \pm z_{0.025}\frac{S}{\sqrt{n}}$. Using $z_{0.025} = 1.96$, $n = 50$, and the summary statistics $\overline{x} = 95.74$ dollars, $s = 24.63$ dollars, the 95% confidence interval for μ (true mean amount contested) is given by $95.74 \pm 1.96\frac{24.63}{\sqrt{50}} = 95.74 \pm 6.83$ or $(88.91, 102.57)$ dollars.

8.29 For large *n*, a 95% confidence interval for μ is given by $\overline{X} \pm z_{0.025}\frac{S}{\sqrt{n}}$. Using $z_{0.025} = 1.96$, $n = 41$, and the summary statistics $\overline{x} = -0.00202$, $s = 0.0234$, the 95% confidence interval for μ is given by

$$0.00202 \pm 1.96\frac{0.0234}{\sqrt{41}} = 0.00202 \pm 0.0072 \text{ or } (-0.00516, 0.00922).$$

8.31 For large *n*, a 99% confidence interval for μ is given by $\overline{X} \pm z_{0.005}\frac{S}{\sqrt{n}}$. Using $z_{0.005} = 2.58$, $n = 40$, and the summary statistics $\overline{x} = 1.715$ centimeters, $s = 5.475$ centimeters, the 99% confidence interval for μ is given by

$$1.715 \pm 2.58\frac{5.475}{\sqrt{40}} = 1.715 \pm 2.233 \text{ or (-0.518, 3.948) centimeters.}$$

8.33 (a) Uncertain since the true mean growth (a population parameter) is not known. Refer to the discussion in the text for a detailed explanation.

(b) The very definition of a confidence interval ensures this statement is true – see property 2 in the text.

8.35 (a) For large *n*, a 95% confidence interval for μ is given by $\overline{X} \pm z_{0.025}\frac{S}{\sqrt{n}}$. Using $z_{0.025} = 1.96$, $n = 40$, and the summary statistics $\overline{x} = 3.56$, $s = 0.05$, the 95% confidence interval for μ (true mean amount of PCBs in the soil) is given by

$$3.56 \pm 1.96\frac{0.05}{\sqrt{40}} = 3.56 \pm 0.155 \text{ or (3.405, 3.715) PCBs.}$$

(b) The sample mean is $\overline{x} = 3.56$ and does indeed lie in this interval. In fact, it is the midpoint of the interval. (This is always true for the *sample* mean.)

(c) Uncertain since the true mean growth (a population parameter) is not known. Refer to the discussion in the text for a detailed explanation.

(d) No. This is an incorrect interpretation of a confidence interval. Refer to the text for a correct interpretation.

8.37 The alternative hypothesis H_1 is the assertion that is to be established; its opposite is the null hypothesis H_0.

(a) Let μ denote the population mean time, in days, to pay a claim. The hypotheses are: $H_0: \mu = 14$, $H_1: \mu < 14$

(b) Let μ denote the population mean amount spent, in dollars. The hypotheses are: $H_0: \mu = 6.50$, $H_1: \mu > 6.50$

(c) Let μ denote the population mean hospital bill, in dollars. The hypotheses are: $H_0: \mu = 7000$, $H_1: \mu < 7000$

(d) Let μ denote the population mean time between purchases, in days. The hypotheses are: $H_0: \mu = 60$, $H_1: \mu \neq 60$

8.39 Retaining H_0 is a correct decision if H_0 is true, while it is an incorrect decision if H_1 is true. A Type II error is incurred when such an incorrect decision is made.

(a) Correct decision if $\mu = 14$ and incorrect decision if $\mu < 14$ (Type II error).

(b) Correct decision if $\mu = 6.50$ and incorrect decision if $\mu > 6.50$ (Type II error).

(c) Correct decision if $\mu = 7000$ and incorrect decision if $\mu < 7000$ (Type II error).

(d) Correct decision if $\mu = 60$ and incorrect decision if $\mu \neq 60$ (Type II error).

8.41 (b) (i) $H_0: \mu = 0.15$, $H_1: \mu < 0.15$

(ii) $Z = \dfrac{\overline{X} - 0.15}{0.085/\sqrt{125}}$

(iii) $R: Z \leq -1.96$

(c) (i) $H_0: \mu = 80$, $H_1: \mu \neq 80$

(ii) $Z = \dfrac{\overline{X} - 80}{8.6/\sqrt{38}}$

(iii) $R: |Z| \geq 2.58$

(d) (i) $H_0: \mu = 0$, $H_1: \mu \neq 0$

(ii) $Z = \dfrac{\overline{X} - 0}{1.23/\sqrt{40}}$

(iii) $R: |Z| \geq 1.88$

8.43 Since the claim is that $\mu > 40$, we formulate the hypotheses:

$$H_0: \mu = 40, \quad H_1: \mu > 40$$

With $n = 70$ and $\sigma = 5.6$, the test statistic (in standardized form) is $Z = \dfrac{\overline{X} - 40}{5.6/\sqrt{70}}$.

(a) $\overline{X} \geq 41.31$ is equivalent to $Z \geq \dfrac{41.31 - 40}{5.6/\sqrt{70}} = 1.96$, or more simply $Z \geq 1.96$.

Since $P[Z \geq 1.96] = 0.025$, the level of significance is $\alpha = 0.025$.

(b) Since $P[Z \geq 1.645] = 0.05$, the rejection region $R: Z \geq 1.645$ has the level of significance $\alpha = 0.05$. Now, since $Z = \dfrac{\overline{X} - 40}{5.6/\sqrt{70}}$, we observe that $Z \geq 1.645$ is equivalent to $\overline{X} \geq 40 + (1.645)\dfrac{5.6}{\sqrt{70}} = 41.10$. Therefore, $c = 41.10$.

8.45 (a) The test statistic remains the same, but this time the rejection region is two-sided. With $\alpha = 0.02$, we have $z_{\alpha/2} = z_{0.01} = 2.33$ and the two-sided rejection region is $R: |Z| \geq 2.33$. Note that the observed value $|z| = 2.26$ is not in this rejection region. Hence, H_0 is not rejected at $\alpha = 0.02$. Furthermore, the *p*-value is $2P[Z \geq 2.26] = 2(0.0119) = 0.0238$.

(b) Since the test statistic was not in the rejection region, it is possible that we made a Type II error, meaning that the mean number of items returned to the store really wasn't 2, but we accepted that it was.

8.47 (a) Reject H_0 since the *p*-value is 0.005, which is less than 0.01.
(b) We could have made a Type I error if H_0 happened to be true.
(c) Prior to sampling, the probability of making a Type I error is 0.01.
(d) If we took more and more samples and ran the same test of hypothesis, about 1% of the time we would make a Type I error (i.e., reject a true H_0).
(e) The *p*-value is 0.005, as reported on the Minitab output.

8.49 (a) Since the claim is that $\mu > 3.5$, we formulate the hypotheses:

$$H_0: \mu = 3.5, \quad H_1: \mu > 3.5$$

Now, to run the test with sample size $n = 40$, we use the test statistic $Z = \dfrac{\overline{X} - 3.5}{S/\sqrt{40}}$.

Since H_1 is right-sided, the rejection region should have the form $R: Z \geq z_\alpha$. With $\alpha = 0.10$, the rejection region is $R: Z \geq 1.28$. The observed value of the test statistic is

$$z = \frac{3.8 - 3.5}{1.2/\sqrt{40}} = 1.58,$$

which is in *R*. Therefore, we reject H_0 at $\alpha = 0.10$. Furthermore, the *p*-value is $P[Z \geq 1.58] = 0.0571$.

(b) We could have made a Type I error, meaning that the average time required to process the rebate was really 3.5 minutes, even though we rejected this claim.

8.51 (a) Let μ be mean battery life. We test the hypotheses:

$$H_0: \mu = 183, \quad H_1: \mu > 183$$

(b) To run the test with sample size $n = 64$, we use the test statistic $Z = \dfrac{\overline{X} - 183}{S / \sqrt{64}}$.

Since H_1 is one-sided, the rejection region should have the form $R: Z \geq z_\alpha$. With $\alpha = 0.05$, the rejection region is $R: Z \geq 1.645$. The observed value of the test statistic is

$$z = \frac{190.5 - 183}{32 / \sqrt{64}} = 1.875,$$

which is in *R*. Therefore, we reject H_0 at $\alpha = 0.05$. Furthermore, the *p*-value is $P[Z \geq 1.875] = 0.0304$.

(c) We could have made a Type I error, meaning that we rejected the fact that the mean battery life was 183 days when in fact it was.

8.53 Let μ denote the population mean hold time (in minutes). Since the claim is that $\mu > 3.0$, we formulate the following hypotheses: $H_0: \mu = 3.0, \quad H_1: \mu > 3.0$

Now, to run the test with sample size $n = 75$, we use the test statistic $Z = \dfrac{\overline{X} - 3.0}{S / \sqrt{75}}$.

Since H_1 is right-sided, the rejection region should have the form $R: Z \geq z_\alpha$. With $\alpha = 0.05$, the rejection region is $R: Z \geq 1.645$. The observed value of the test statistic is

$$z = \frac{3.4 - 3.0}{2.4 / \sqrt{75}} = 1.44,$$

which is not in *R*. Therefore, we do not reject H_0 at $\alpha = 0.05$. Furthermore, the *p*-value is $P[Z \geq 1.44] = 0.0749$. Since this is not small, support for the claim is weak.

8.55 Let μ denote the true mean BOD. Since the intent is to establish that μ is different from 3000, we formulate the following hypotheses:

$$H_0: \mu = 3000, \quad H_1: \mu \neq 3000$$

Now, to run the test with sample size $n = 43$, we use the test statistic $Z = \dfrac{\overline{X} - 3000}{S / \sqrt{43}}$

Since H_1 is two-sided, the rejection region should have the form $R: |Z| \geq z_{\alpha/2}$. With $\alpha = 0.05$, $z_{0.025} = 1.96$, so that the rejection region is $R: |Z| \geq 1.96$. The observed value of the test statistic is

$$z = \frac{3246 - 3000}{757 / \sqrt{43}} = 2.13,$$

which is in R. Therefore, we reject H_0 at $\alpha = 0.05$. Moreover, the p-value is $P[Z > 2.13] = 0.0166$. This means that there is strong evidence that the BOD is significantly off the target.

8.57 (a) (i) p = proportion of adult population for which reading is a favorite leisure time activity.

(ii) Point estimate $\hat{p} = \frac{295}{986} = 0.299$

(iii) Estimated S.E. = $\sqrt{\frac{\hat{p}\hat{q}}{n}} = \sqrt{\frac{(0.299)(0.701)}{986}} = 0.0146$

So, the 95% error margin = 1.96(Estimated S.E.) = $1.96(0.0146) = 0.0286$.

(b) (i) p = proportion of pet owners who revealed that they buy their pets holiday presents.

(ii) Point estimate $\hat{p} = \frac{293}{440} = 0.666$

(iii) Estimated S.E. = $\sqrt{\frac{\hat{p}\hat{q}}{n}} = \sqrt{\frac{(0.666)(0.334)}{440}} = 0.022$

So, the 95% error margin = 1.96(Estimated S.E.) = $1.96(0.022) = 0.043$.

8.59 (a) We have $p = 0.3,\ q = 0.7,\ d = 0.03,\ \alpha = 0.10$. So, $z_{\alpha/2} = z_{0.05} = 1.645$. We calculate $n = (0.3)(0.7)\left[\frac{1.645}{0.03}\right]^2 = 631.40$. So, the required sample size is 632.

(b) Since p is unknown, the conservative bound on n yields $\frac{1}{4}\left[\frac{1.645}{0.03}\right]^2 = 751.67$, so that the required sample size in this instance is 752.

8.61 (a) Denote p_M = population proportion of ERS calls involving serious mechanical problems. Since there were 849 calls involving serious mechanical problems out of $n = 2927$ ERS calls, the point estimate of p_M is $\hat{p}_M = \frac{849}{2927} = 0.290$. Also, the 95% error margin is $1.96\sqrt{\frac{(0.29)(0.71)}{2927}} = 0.016$.

(b) Denote p_S = population proportion of ERS calls involving starting problems. Since there were 1499 calls involving starting problems out of $n = 2927$ ERS calls, the point estimate of p_S is $\hat{p}_S = \frac{1499}{2927} = 0.512$, and so $\hat{q}_S = 1 - 0.512 = 0.488$. The 98% confidence interval for p_S is then given by

$$\hat{p}_S \pm 2.33\sqrt{\frac{\hat{p}_S\hat{q}_S}{n}} = 0.512 \pm 2.33\sqrt{\frac{(0.512)(0.488)}{2927}} = 0.512 \pm 0.0215$$

or (0.491, 0.534).

8.63 Let $p =$ population proportion of all drive-up window purchases made with a major credit card.

(a) The sample proportion is $\hat{p} = \frac{27}{94} = 0.287$.

(b) Estimated standard error is $\sqrt{\frac{\hat{p}\hat{q}}{n}} = \sqrt{\frac{(0.287)(0.713)}{94}} = 0.047$

(c) The 98% confidence interval for p is given by

$$\hat{p} \pm z_{0.01}\sqrt{\frac{\hat{p}\hat{q}}{n}} = 0.287 \pm 2.33(0.047) = 0.287 \pm 0.110$$

or (0.177, 0.397). That is, (17.7%, 39.7%).

8.65 (a) Let p be the proportion of students that hold a part time job. The hypotheses are: $H_0: p = 0.26, \quad H_1: p > 0.26$

(b) Let p be the proportion of subscribers that have complaints against the cable company. The hypotheses are: $H_0: p = 0.13, \quad H_1: p < 0.13$

(c) Let p be the proportion of subscribers that have complaints against the cable company. The hypotheses are: $H_0: p = 0.13, \quad H_1: p > 0.13$

(b) Let p be the proportion of boards that would break under the standard load. The hypotheses are: $H_0: p = 0.05, \quad H_1: p < 0.05$

8.67 (b) (i) $H_0: p = 0.75, \quad H_1: p > 0.75$

(ii) $Z = \frac{\hat{p} - 0.75}{\sqrt{\frac{(0.75)(0.25)}{228}}} = \frac{\hat{p} - 0.75}{0.0287}$

(iii) Since $\alpha = 0.02$, $z_{0.02} = 2.05$ and H_1 is right-sided, the rejection region is given by $R: Z \geq 2.05$.

(c) (i) $H_0: p = 0.60, \quad H_1: p \neq 0.60$

(ii) $Z = \frac{\hat{p} - 0.60}{\sqrt{\frac{(0.60)(0.40)}{77}}} = \frac{\hat{p} - 0.60}{0.0558}$

(iii) Since $\alpha = 0.02$, $\alpha/2 = 0.01$, $z_{0.01} = 2.33$ and H_1 is two-sided, the rejection region is given by $R: |Z| \geq z_{\alpha/2} = 2.33$.

(d) (i) $H_0: p = 0.56, \quad H_1: p < 0.56$

(ii) $Z = \frac{\hat{p} - 0.56}{\sqrt{\frac{(0.56)(0.44)}{86}}} = \frac{\hat{p} - 0.56}{0.0535}$

(iii) Since $\alpha = 0.10$, $z_{0.10} = 1.28$ and H_1 is left-sided, the rejection region is given by $R: Z \leq -1.28$.

8.69 Let p = proportion of all freshmen dorms that have a poster of a rock group hanging.

(a) Test the hypotheses $H_0 : p = 0.30, \quad H_1 : p > 0.30$

(b) Since H_1 is one-sided, the rejection region $R : Z \geq z_{\alpha}$, and since $\alpha = 0.05$, this becomes $R : Z \geq 1.645$.

(c) We would reject H_0 in favor of H_1. So, a type I error could be made.

(d) Test statistic: $Z = \dfrac{\hat{p} - 0.30}{\sqrt{\frac{(0.70)(0.30)}{60}}} = \dfrac{\hat{p} - 0.30}{0.059}$

If $x = 25$, then $\hat{p} = \dfrac{25}{60} = 0.417$. Then, the test statistic value is given by

$Z = \dfrac{0.417 - 0.30}{0.059} = 1.983$, which is inside R. Hence, we reject H_0 at $\alpha = 0.05$.

(e) Since we are rejecting H_0, we could have made a Type I error, meaning that the true proportion of dorms with posters of a rock group hanging really is 0.30.

8.71 (a) Since the intent is to establish that $p < 0.5$, we formulate the following hypotheses: $H_0 : p = 0.5, \quad H_1 : p < 0.5$

(b) The sample proportion of support is $\hat{p} = \frac{228}{500} = 0.456$. Since $\alpha = 0.05$, $z_{0.05} = 1.645$ and H_1 is left-sided, the rejection region is $R : Z \leq -1.645$. The test statistic is calculated to be $Z = \dfrac{\frac{228}{500} - 0.5}{\sqrt{\frac{(0.5)(0.5)}{500}}} = -1.97$, which is in R. Hence, H_0 is rejected at $\alpha = 0.05$ and the claim is supported. Furthermore, the p-value is $P[Z \leq -1.97] = 0.0244$, so that H_0 would still be rejected with α as small as 0.0244.

8.73 Let p denote the population proportion of students who sometimes use cell phones while driving. Since the intent is to establish that $p > 0.75$, we formulate the following hypotheses:

$$H_0 : p = 0.75, \quad H_1 : p > 0.75$$

Since H_1 is right-sided, the rejection region is of the form $R : Z \geq z_{\alpha} = 1.645$. The sample proportion is $\hat{p} = 0.78$. The test statistic is calculated to be

$$Z = \frac{\hat{p} - 0.75}{\sqrt{\frac{(0.75)(0.25)}{5000}}} = \frac{0.78 - 0.75}{0.006} \approx 5.$$

The is certainly in the rejection region, so that there is very strong evidence in support of our claim.

8.75 (a) We have $n = 505$ and $\hat{p} = \frac{258}{505} = 0.5109$. So, a 90% confidence interval for p is given by:

$$\hat{p} \pm z_{0.05}\sqrt{\frac{\hat{p}\hat{q}}{n}} = 0.5109 \pm 1.645\sqrt{\frac{(0.5109)(0.4881)}{505}} = 0.5109 \pm 0.0366$$

or (0.4743, 0.5475).

(b) The unknown total number of customers who would rate the service as being excellent is 8200p. The lower endpoint of the 90% confidence interval for the total is then 8200(0.4743) = 3889, and the upper endpoint is 8200(0.5475) = 4490.

8.77 We have $n = 42$ and $\hat{p} = \frac{30}{42} = 0.7143$. So, a 95% confidence interval for p is given by:

$$\hat{p} \pm z_{0.025}\sqrt{\frac{\hat{p}\hat{q}}{n}} = 0.7143 \pm 1.96\sqrt{\frac{(0.7143)(0.2857)}{42}} = 0.7143 \pm 0.1366$$

or (0.58, 0.85). That is, (58%, 85%).

8.79 Recall that the point estimate of μ is $\overline{x} = \dfrac{\sum x_i}{n}$ and the estimated standard error is $\frac{s}{\sqrt{n}}$, where $s^2 = \dfrac{\sum(x_i - \overline{x})^2}{n-1}$.

(a) $\overline{x} = \dfrac{3230.84}{38} = 85.02$

(b) $\text{S.E.} = \dfrac{s}{\sqrt{n}} = \dfrac{\sqrt{\dfrac{2028.35}{37}}}{\sqrt{38}} = 1.201$

(c) 95% error margin is $z_{0.025}\frac{\sigma}{\sqrt{n}} = (1.96)(1.201) = 2.354$

8.81 Recall that the standard error of $\overline{X}$ is $\dfrac{\sigma}{\sqrt{n}}$, where n is the sample size.

(a) Since $\dfrac{1}{2}\dfrac{\sigma}{\sqrt{n}} = \dfrac{\sigma}{\sqrt{4n}}$, we need a sample of size 4n. Therefore, we must increase the sample size by a factor of 4.

(b) Since $\dfrac{1}{4}\dfrac{\sigma}{\sqrt{n}} = \dfrac{\sigma}{\sqrt{16n}}$, we need a sample of size 16n. Therefore, we must increase the sample size by a factor of 16.

8.83 (a) The population mean μ is estimated by $\overline{x} = 126.9$.

Estimated $\text{S.E.} = \dfrac{s}{\sqrt{n}} = \dfrac{10.5}{\sqrt{55}} = 1.416$. So, the approximate 95.4% error margin is given by $2\,(\text{Estimated S.E.}) = 2(1.416) \approx 2.8$,

(b) Observe that $1-\alpha = 0.90$ implies $\alpha = 0.10$, so that $z_{\alpha/2} = z_{0.05} = 1.645$. A 90% confidence interval for μ is calculated as follows:

$$\bar{x} \pm 1.645\frac{s}{\sqrt{n}} = 126.9 \pm 1.645(1.416) = 126.9 \pm 2.329 \text{ or } (124.571, 129.228).$$

8.85 (a) Correct

(b) We will never know whether this particular interval covers the true mean μ. This is a single realization of the random interval $\overline{X} \pm 1.645\frac{S}{\sqrt{n}}$. In repeated sampling, about 90% of such intervals will cover the true mean μ.

8.87 The alternative hypothesis H_1 is the assertion that is to be established; its opposite is the null hypothesis H_0.

(a) Let μ denote the population mean mileage. The hypotheses are: $H_0: \mu = 50, \quad H_1: \mu < 50$

(b) Let μ denote the population mean number of pages per transmission. The hypotheses are: $H_0: \mu = 3.4, \quad H_1: \mu > 3.4$

(c) Let p denote the probability of success with the method. The hypotheses are: $H_0: p = .50, \quad H_1: p > .50$

(d) Let μ denote the mean fill. The hypotheses are: $H_0: \mu = 16, \quad H_1: \mu \neq 16$

(e) Let μ denote the mean percent fat content. The hypotheses are: $H_0: p = 0.04, \quad H_1: p > 0.04$

8.89 Let μ denote the mean length of time that a 7 oz. tube of toothpaste lasts.

(a) We test the hypotheses: $H_0: \mu = 30.5, \quad H_1: \mu > 30.5$

(b) The test statistic is $Z = \dfrac{\overline{X} - 30.5}{s/\sqrt{n}}$.

(c) Since H_1 is one-sided, the rejection region $R: Z \geq z_\alpha$

(d) Using $\bar{x} = 32.3$, $s = 6.2$, $n = 75$, we see that the test statistic is $Z = \dfrac{32.3 - 30.5}{6.2/\sqrt{75}} = 2.514$. For $\alpha = 0.10$, the rejection region is $R: Z \geq z_{0.10} = 1.28$. Since the test statistic value is in R, we reject H_0 at $\alpha = 0.10$.

(e) The associated p-value is $P(Z > 2.514) = 0.00597$.

(f) Since we rejected H_0, we could have made a Type I error in that we rejected the fact that there are 9.1 words per sentence, on average, when this is in fact the case.

8.91 (a) Reject H_0 since the p-value is 0.014, which is less than $\alpha = 0.03$.

(b) To test the hypotheses: $H_0: \mu = 84 \quad H_1: \mu < 84$

we use the test statistic $\dfrac{\overline{X} - 84}{S/\sqrt{n}}$ (the observed value of which, from the printout, is -2.45). Since H_1 is left-sided, the rejection region should have the form $R: Z \le z_\alpha$. Since $z_{0.01} = -2.33$, the rejection region is $R: Z \le -2.33$. Since the value of the test statistic, namely -2.45, is in R, we reject H_0 at $\alpha = 0.01$.

8.93 (a) $E[X] = 0(0.5) + 1(0.3) + 2(0.2) = 0.7$

(b) Since $E[X^2] = 0^2(0.5) + 1^2(0.3) + 2^2(0.2) = 1.1$, we observe that

$\text{var}[X] = E[X^2] - (E[X])^2 = 1.1 - (0.7)^2 = 0.61$, and so $\sigma = \sqrt{0.61} = 0.781$

(c) To test $H_0: \mu = 0.7$ versus $H_1: \mu > 0.7$, we use the test statistic

$Z = \dfrac{\overline{X} - 0.7}{S/\sqrt{n}}$ with $n = 64$ and we will use the standard deviation from the new sample because the distribution of sales may have changed. With $\alpha = 0.05$, the rejection region is $R: Z \ge 1.645$. The observed value of the test statistic is

$$z = \frac{0.84 - 0.7}{0.4/\sqrt{64}} = 2.80,$$

which is in R. Therefore, we reject H_0 at $\alpha = 0.05$. Furthermore, the p-value is $P[Z \ge 2.80] = 0.0026$. So, the evidence in favor of increased mean sales is very strong.

8.95 (a) The point estimate of the proportion of passport holders is $\hat{p} = \frac{64}{84} = 0.762$.

(b) The 95% error margin = 1.96(Estimated S.E.), which is given by

$$1.96\sqrt{\frac{\hat{p}\hat{q}}{n}} = 1.96\sqrt{\frac{(0.762)(0.238)}{84}} = 0.046.$$

8.97 We have $n = 625$ and $\hat{p} = \frac{139}{625} = 0.2224$, so that $\hat{q} = 0.7776$. We also are given that $\alpha = 0.05$ (so that $z_{\alpha/2} = z_{0.025} = 1.96$) and. Thus, the 95% confidence interval is

$$\hat{p} \pm z_{0.025}\sqrt{\frac{\hat{p}\hat{q}}{n}} = 0.2224 \pm 1.96\sqrt{\frac{(0.2224)(0.7776)}{625}} = 0.2224 \pm 0.0333 \quad \text{or} \quad (0.189, 0.256).$$

That is, (18.9%, 25.6%).

8.99 Let p = proportion of all flights from Chicago to Austin that arrive late.

(a) We test the hypotheses: $H_0: p = 0.10, \quad H_1: p > 0.10$

(b) Denoting the sample proportion by $\hat{p}$, the test statistic is $Z = \dfrac{\hat{p} - 0.10}{\sqrt{\frac{\hat{p}\hat{q}}{n}}}$

(c) Since H_1 is one-sided, the rejection region with $\alpha = 0.05$ is $R: Z \geq 1.645$.

(d) Using $\hat{p} = \frac{22}{152} = 0.145$ and $n = 152$, the value of the test statistic is $Z = \dfrac{0.145 - 0.10}{\sqrt{\frac{(0.145)(0.855)}{152}}} = 1.576$. Since this is in not R, we do not reject H_0 at $\alpha = 0.05$.

(e) The associated p-value is $P(Z > 1.576) = 0.0577$.

(f) Since we did not reject H_0, we could have made a Type II error, meaning that the claim is valid when we assert it is not.

8.101 (a) Let p denote the population proportion of plants that are of the dwarf variety. Since the intent is to establish that $p \neq 0.8$, we formulate the following hypotheses: $H_0: p = 0.8,\quad H_1: p \neq 0.8$ The test statistic for a sample of size $n = 200$ is $Z = \dfrac{\hat{p} - 0.8}{\sqrt{\frac{(0.8)(0.2)}{200}}} = \dfrac{\hat{p} - 0.8}{0.0283}$. With $\alpha = 0.05$, $z_{0.05/2} = 1.96$ and H_1 is two-sided, the rejection region is $R: |Z| \geq 1.96$. The observed sample proportion is $\hat{p} = \frac{136}{200} = 0.68$, and the test statistic has the value $Z = \dfrac{0.68 - 0.8}{0.0283} = -4.24$, which is in R. Hence, H_0 is rejected at $\alpha = 0.05$. Furthermore, the associated p-value is given by

$$P[Z \leq -4.24] + P[Z \geq 4.24] < 0.0001,$$

which is extremely small. As such, a contradiction of the genetic model is strongly indicated.

(b) A 95% confidence interval for p is given by

$$\hat{p} \pm 1.96\sqrt{\frac{\hat{p}\hat{q}}{n}} = 0.68 \pm 1.96\sqrt{\frac{(0.68)(0.32)}{200}} = 0.68 \pm 0.0555 \quad \text{or} \quad (0.6245,\ 0.7355).$$

8.103 The 95% confidence interval is

$$\bar{x} \pm z_{0.025}\frac{s}{\sqrt{n}} = 19.22 \pm 1.96\left(\frac{17.66}{\sqrt{65}}\right) \quad \text{or} \quad (14.927,\ 23.515).$$

8.105 (a) A 99% confidence interval will be larger than the 95% confidence interval given in the printout. In fact, the greater the confidence <u>level</u>, the larger the confidence <u>interval</u> will be for a given sample size. To verify this, observe that

$$\bar{x} \pm z_{0.01/2}\frac{s}{\sqrt{n}} = 2.8157 \pm 2.58\frac{0.4840}{\sqrt{35}} = 2.8157 \pm 0.2110, \quad \text{or} \quad (2.605,\ 3.026).$$

(b) To test $H_0: \mu = 2.6$ versus $H_1: \mu > 2.6$, we use the test statistic $Z = \dfrac{\overline{X} - 2.6}{0.484/\sqrt{35}}$. With $\alpha = 0.05$, since H_1 is right-sided, the rejection region is $R: Z \geq 1.645$. The observed value of the test statistic is

$$z = \frac{2.8157 - 2.6}{0.484/\sqrt{35}} = 2.637,$$

which is in *R*. Therefore, we reject H_0 at $\alpha = 0.05$.

8.107 Enter the data into column C2 in a Minitab worksheet.

(a) Run a 1-sample *Z*-test since the population standard deviation is unknown. The output is as follows.

Z Confidence Intervals

```
Variable      N        Mean     StDev   SE Mean        97.0 % CI
C2           40      1.7150    0.4748    0.0751   (1.5459, 1.8841)
```

So, the 97% confidence interval is (1.5459, 1.8841).

(b) Testing the hypotheses $H_0: \mu = 1.9$ versus $H_1: \mu \neq 1.9$ using Minitab yields the following output:

Z-Test of the Mean

```
Test of mu = 1.9000 vs mu not = 1.9000

Variable    N       Mean     StDev       SE Mean        T        P
C2           40     1.7150    0.4748      0.0751    -2.46    0.018
```

Since the *p*-value is 0.018, we reject H_0 at $\alpha = 0.03$.

8.109 Enter the data into column C4 in a Minitab worksheet. Run a 1-sample *Z*-test since the population standard deviation is unknown. The output is as follows.

Z Confidence Intervals

```
Variable      N        Mean     StDev    SE Mean        90.0 % CI
C4           28      110.39     23.18     4.38    (102.93, 117.85)
```

So, the 90% confidence interval is (102.93, 117.85).

8.111 (a) Enter the data into column C6 in a Minitab worksheet. Run a 1-sample *Z*-test since the population standard deviation is unknown. The output is as follows.

Z Confidence Intervals

```
Variable      N        Mean     StDev      SE Mean        95.0 % CI
C6          151       4.072     9.065      0.738    ( 2.614, 5.530)
```

So, the 95% confidence interval is (2.614, 5.530).

(b) The histogram is as follows. Note that the histogram has a long tail to the right (much more extreme than in Exercise 8.108).

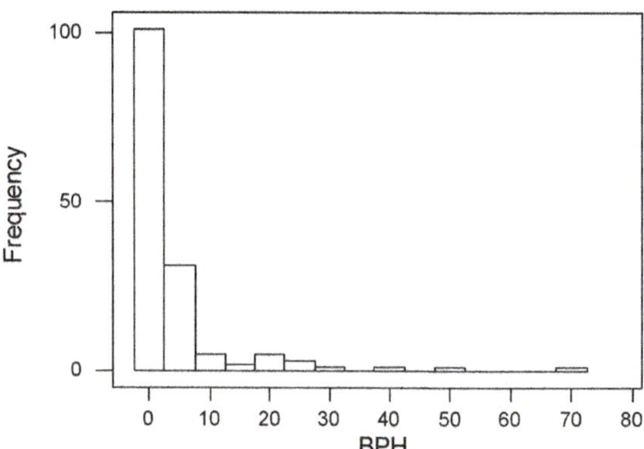

Chapter 9

SMALL-SAMPLE INFERENCES FOR NORMAL POPULATIONS

9.1 (a) $t_{0.05} = 1.943$

(b) $-t_{0.025} = -2.228$

(c) $-t_{0.01} = -2.821$

(d) $t_{0.10} = 1.350$

9.3 (a) 90th percentile of $t = t_{0.10} = 1.397$

(b) 99th percentile of $t = t_{0.01} = 3.365$

(c) 5th percentile of $t = -t_{0.05} = -1.717$

(d) upper quartile $= t_{0.25} = 0.689$

lower quartile $= -t_{0.25} = -0.689$

9.5 (a) Since the area to the right of b is $1-0.95 = 0.05$, b is the upper 0.05 point of the t distribution with d.f. = 7. From the t-table we find that $b = t_{0.05} = 1.895$.

(b) Since $P[T > b] = 0.025$, from the t-table with d.f. = 16, we find that $b = t_{0.025} = 2.120$.

(c) $b = t_{0.01} = 3.365$

(d) Since $P[T > b] = 0.01$, b is the upper 0.01 point of the t-distribution, namely $b = t_{0.01} = 2.718$.

9.7 (b) In the t-table with d.f. = 16, we look for the percentage points that are close to the given number 1.9. We find that $t_{0.05} = 1.746$ and $t_{0.025} = 2.120$. Because 1.9 lies between $t_{0.025}$ and $t_{0.05}$, the probability $P[T > 1.9]$ must lie between 0.025 and 0.05.

(c) The number 1.5 is between $t_{0.10} = 1.363$ and $t_{0.05} = 1.796$, and $P[T < -1.5] = P[T > 1.5]$. Thus, the probability $P[T < -1.5]$ must lie between 0.05 and 0.10.

(d) The number 1.9 is between $t_{0.05} = 1.812$ and $t_{0.025} = 2.228$, and $P[|T| > 1.9] = 2P[T > 1.9]$. Thus, the probability $P[|T| > 1.9]$ must lie between 2(0.025) and 2(0.05); that is, between 0.05 and 0.10.

(e) With d.f. = 17, the number 2.8 is between $t_{0.01} = 2.567$ and $t_{0.005} = 2.898$, so the probability $P[T > 2.8]$ must lie between 0.005 and 0.01. Because $P[|T| < 2.8] = 1 - 2P[T > 2.8]$, this probability must lie between $1 - 2(0.005) = 0.99$ and $1 - 2(0.01) = 0.98$.

9.9 (a) A 98% confidence interval for μ is given by $\overline{X} \pm t_{0.01}\dfrac{S}{\sqrt{n}}$, where d.f. $= n-1$.

For d.f. = 19, we find that $t_{0.01} = 2.539$. In such case, the confidence interval is

$$76.1 \pm 2.539\left(\frac{5.9}{\sqrt{20}}\right) = 76.1 \pm 3.35 \quad \text{or} \quad (72.75,\ 79.45).$$

(b) Center is $\overline{x} = 76.1$. Length is 2(3.35) = 6.70.

(c) Usually different since the length of the interval depends on the sample standard deviation S which varies from sample to sample.

9.11 Assume a normal population. A 95% confidence interval for the true mean μ is given by $\overline{X} \pm t_{0.025}\dfrac{S}{\sqrt{n}}$, where d.f. $= n-1$. For sample size $n = 20$, we have d.f. $= n-1 = 19$ and $t_{0.025} = 2.093$. From the sample data, we calculate $\overline{x} = 137.60$ and $s = 20.143$. The 95% confidence interval for μ is then given by

$$137.60 \pm 2.093\left(\frac{20.143}{\sqrt{20}}\right) = 137.60 \pm 9.43 \quad \text{or} \quad (128.17,\ 147.03).$$

9.13 A 95% confidence interval for the true mean μ is given by $\overline{X} \pm t_{0.025}\dfrac{S}{\sqrt{n}}$, where d.f. $= n-1$. For sample size $n = 18$, we have d.f. $= n-1 = 17$ and $t_{0.025} = 2.110$. From the sample data, we know $\overline{x} = 3.6$ and $s = 0.8$. The 95% confidence interval for μ is then given by

$$3.6 \pm 2.110\left(\frac{0.8}{\sqrt{18}}\right) = 3.6 \pm 0.398 \quad \text{or} \quad (3.202,\ 3.998)$$

When forming this confidence interval, we assumed the underlying population was normally distributed.

9.15 Assume a normal population. A 95% confidence interval for the true mean μ is given by $\overline{X} \pm t_{0.025}\frac{S}{\sqrt{n}}$, where d.f. = $n-1$. In this case, $n=12$, so that d.f. $= n-1 = 11$ and $t_{0.025} = 2.201$.

(a) A calculation of $\overline{x} \pm 2.201\frac{s}{\sqrt{12}}$ has yielded the result (38.6, 46.2). This interval has

$$\text{center} = \overline{x} = \frac{38.6+46.2}{2} = 42.4, \quad \text{half-width} = 2.201\frac{s}{\sqrt{12}} = 42.4 - 38.6 = 3.8.$$

From the last relation, we see that $s = 3.8\left(\frac{\sqrt{12}}{2.201}\right) = 5.98$. (b) A 98% confidence interval for the true mean μ is given by $\overline{X} \pm t_{0.01}\frac{S}{\sqrt{n}}$, where d.f. = $n-1$. For sample size $n=12$, we have d.f. $= n-1 = 11$ and $t_{0.01} = 2.718$. From part (a), we have $\overline{x} = 42.4$ and $s = 5.98$. The 98% confidence interval for μ is then given by

$$42.4 \pm 2.718\left(\frac{5.98}{\sqrt{12}}\right) = 42.4 \pm 4.7 \quad \text{or} \quad (37.7, 47.1).$$

9.17 Assume a normal population. A 99% confidence interval for the true mean μ is given by $\overline{X} \pm t_{0.005}\frac{S}{\sqrt{n}}$, where d.f. = $n-1$. For sample size $n=23$, we have d.f. $= n-1 = 22$ and $t_{0.005} = 2.819$. From the sample data, we have $\overline{x} = 5.483$ and $s = 0.1904$. So, the 99% confidence interval for μ is then given by

$$5.483 \pm 2.819\left(\frac{0.1904}{\sqrt{23}}\right) = 5.483 \pm 0.111 \quad \text{or} \quad (5.372, 5.594).$$

9.19 We are to test the hypotheses $H_0: \mu = 3.8$ versus $H_1: \mu < 3.8$ with $\alpha = 0.05$. The test statistic is $T = \frac{\overline{X} - 3.8}{S/\sqrt{14}}$. For sample size $n = 14$, we have d.f. $= n-1 = 13$ and $t_{0.05} = 1.771$. Since H_1 is one-sided, the rejection region is $R: T < -1.771$. From the sample data, we have $\overline{x} = 3.363$ and $s = 3.354$. The value of the observed t is then $t = \frac{3.363 - 3.8}{0.507/\sqrt{14}} = -3.225$, which lies in R. Hence, H_0 is rejected.

9.21 (a) Uncertain since the true mean length (a population parameter) is not known. Refer to the text for a detailed explanation.
(b) The very definition of a confidence interval ensures this statement is true – see property 2 in the text.

9.23 (a) Assume that the mound diameters are normally distributed. Here, the conjecture is that μ is larger than 21 feet, so we formulate the hypotheses:

$$H_0 : \mu = 21 \text{ versus } H_1 : \mu > 21$$

The test statistic is $T = \dfrac{\overline{X} - 21}{S/\sqrt{13}}$. For sample size $n = 13$ and $\alpha = 0.01$, we have d.f. $= n-1 = 12$ and $t_{0.01} = 2.681$. Since H_1 is right-sided, the rejection region is $R: T \geq 2.681$. From the sample data, we have $\overline{x} = 26.62$ and $s = 6.56$. The value of the observed t is then $t = \dfrac{26.62 - 21}{6.56/\sqrt{13}} = 3.09$, which lies in R. Hence, H_0 is rejected at $\alpha = 0.01$. Furthermore, the associated p-value is $P[T \geq 3.09] = 0.0047$, so there is strong evidence in support of the conjecture.

(b) A 90% confidence interval for the true mean μ is given by $\overline{X} \pm t_{0.05} \dfrac{S}{\sqrt{n}}$, where d.f. $= n-1$. For sample size $n = 13$, we have d.f. $= n-1 = 12$ and $t_{0.05} = 1.782$. From the sample data, we have $\overline{x} = 26.62$ feet and $s = 6.56$ feet. So, the 90% confidence interval for μ is then given by

$$26.62 \pm 1.782\left(\frac{6.56}{\sqrt{13}}\right) = 26.62 \pm 3.24 \text{ or } (23.38, 29.86) \text{ feet.}$$

9.25 Assume that the data are normally distributed. Here, the conjecture is that μ is larger than 128 mm, so we formulate the hypotheses:

$$H_0 : \mu = 128 \text{ versus } H_1 : \mu > 128$$

The test statistic is $T = \dfrac{\overline{X} - 128}{S/\sqrt{20}}$. For sample size $n = 20$ and $\alpha = 0.05$, we have d.f. $= n-1 = 19$ and $t_{0.05} = 1.729$. Since H_1 is right-sided, the rejection region is $R: T \geq 1.729$. From the sample data, we have $\overline{x} = 137.60$ and $s = 20.143$. The value of the observed t is then $t = \dfrac{137.60 - 128}{20.143/\sqrt{20}} = 2.13$, which lies in R. Hence, H_0 is rejected at $\alpha = 0.05$. Furthermore, the associated p-value is $P[T \geq 2.13] = 0.023$, which again is strong evidence in support of the conjecture.

9.27 Assume a normal population. We are to test the hypotheses:

$$H_0 : \mu = 83 \text{ versus } H_1 : \mu \neq 83$$

The test statistic is $T = \dfrac{\overline{X} - 83}{S/\sqrt{8}}$. For sample size $n = 8$ and $\alpha = 0.05$, we have d.f. $= n-1 = 7$ and $t_{0.025} = 2.365$. Since H_1 is two-sided, the rejection region is $R: |T| \geq 2.365$. From the sample data, we have $\overline{x} = 73.87$ and $s = 10.063$. The value of the observed t is then $t = \dfrac{73.87-83}{10.063/\sqrt{8}} = -2.566$, which lies in R. Hence, H_0 is rejected at $\alpha = 0.05$. The associated p-value is $P[|T| \geq 2.566] = 0.0372$.

9.29 (a) Assume that the mean drying time data are normally distributed. Here, we wish to establish a decrease in the mean drying time, so we formulate the hypotheses:

$$H_0 : \mu = 90 \text{ versus } H_1 : \mu < 90$$

The test statistic is $T = \dfrac{\overline{X}-90}{S/\sqrt{15}}$. For sample size $n = 15$ and $\alpha = 0.05$, we have d.f. $= n-1 = 14$ and $-t_{0.05} = -1.761$. Since H_1 is left-sided, the rejection region is $R: T \leq -1.761$. From the sample data, we have $\bar{x} = 86$ and $s = 4.5$. The value of the observed t is then $t = \dfrac{86-90}{4.5/\sqrt{15}} = -3.44$, which lies in R. Hence, H_0 is rejected at $\alpha = 0.05$. Furthermore, scanning the t-table for d.f. = 14, we find that the claim would also be rejected at $\alpha = 0.005$ since $t_{0.005} = 2.977$. In fact, the p-value is approximately 0.002. This extremely small p-value lends a strong support for H_1.

(b) A 98% confidence interval for the mean drying time μ is given by $\overline{X} \pm t_{0.01}\dfrac{S}{\sqrt{n}}$, where d.f. = $n-1$. For sample size $n = 15$, we have d.f. $= n-1 = 14$ and $t_{0.01} = 2.624$. From the sample data, we have $\bar{x} = 86$ and $s = 4.5$. So, the 98% confidence interval for μ is then given by

$$86 \pm 2.624\left(\frac{4.5}{\sqrt{15}}\right) = 86 \pm 3.05 \text{ or } (82.95,\ 89.05).$$

(c) We assumed that the underlying population was normal.

9.31 (a) Reject H_0 since the p-value is 0.022, which is less than $\alpha = 0.05$.
(b) We could have made a Type I error (if H_0 happened to be true).
(c) Prior to sampling, the probability of making a Type I error is 0.05.
(d) If we took more and more samples and ran the same test of hypothesis, about 5% of the time we would make a Type I error (i.e., reject a true H_0).

9.33 (a) The value $\mu_0 = 81$ is outside the 90% confidence interval (67.13, 80.62). Therefore, the null hypothesis $H_0 : \mu_0 = 81$ is rejected at $\alpha = 0.10$.
(b) The value $\mu_0 = 69$ lies inside the 90% confidence interval (67.13, 80.62). Therefore, the null hypothesis $H_0 : \mu_0 = 69$ is not rejected at $\alpha = 0.10$.

9.35 Assume a normal population.

(a) A 95% confidence interval for μ is given by $\overline{X} \pm t_{0.025}\frac{S}{\sqrt{n}}$, where d.f. = $n-1$.

For sample size $n=8$, we have d.f. $= n-1=7$ and $t_{0.025}=2.365$. From the sample data, we have $\overline{x}=6.78$ and $s=6.58$. So, the 95% confidence interval for μ is then given by

$$6.78 \pm 2.365\left(\frac{6.58}{\sqrt{8}}\right) = 6.78 \pm 5.50 \quad \text{or} \quad (1.28,\ 12.28).$$

(b) The value $\mu_0 = 15$ is outside the 95% confidence interval. Therefore, the null hypothesis $H_0 : \mu_0 = 15$ is rejected at $\alpha = 0.05$.

(c) Here, we wish to test the hypotheses: $H_0 : \mu = 15$ versus $H_1 : \mu \neq 15$

The test statistic is $T = \frac{\overline{X}-15}{S/\sqrt{8}}$. For sample size $n=8$ and $\alpha = 0.05$, we have d.f. $= n-1=7$ and $t_{0.025}=2.365$. Since H_1 is two-sided, the rejection region is $R: |T| \geq 2.365$. From the sample data, we have $\overline{x}=6.78$ and $s=6.58$. The value of the observed t is then $t = \frac{6.78-15}{6.58/\sqrt{8}} = -3.53$, which lies in R. Hence, H_0 is rejected at $\alpha = 0.05$. This confirms the conclusion in part (b).

9.37 The acceptance region of the $\alpha = 0.05$ test is $-1.96 \leq \frac{\overline{X}-\mu_0}{S/\sqrt{n}} \leq 1.96$. A rearrangement of these inequalities subsequently yields

$$\overline{X} - 1.96\frac{S}{\sqrt{n}} \leq \mu_0 \leq \overline{X} + 1.96\frac{S}{\sqrt{n}}.$$

Thus, any μ_0 that lies in the interval $\overline{X} \pm 1.96\frac{S}{\sqrt{n}}$ will not be rejected at $\alpha = 0.05$. This interval is precisely the 95% confidence interval for μ.

9.39 (a) Since the area to its left is $1-0.10=0.90$, it is the 90th percentile $\chi^2_{0.10} = 22.31$.

(b) The 5th percentile is $\chi^2_{0.95} = 13.85$.

(c) $\chi^2_{0.10} = 51.81$

(d) $\chi^2_{0.95} = 2.73$

9.41 (a) The sample standard deviation is computed as follows:

$$S^2 = \tfrac{1}{n-1}\left(\sum x^2 - \tfrac{\left(\sum x\right)^2}{n}\right) = \tfrac{1}{14}\left(103.1631 - \tfrac{37.71^2}{15}\right) = 0.597$$

$$S = \sqrt{0.597} \approx 0.7727$$

(b) For $\text{d.f.} = n-1 = 14$, we have $\chi^2_{0.025} = 26.12$ and $\chi^2_{0.975} = 5.62$. The general form for a 95% confidence interval for σ is given by $\left(S\sqrt{\frac{n-1}{\chi^2_{\alpha/2}}}, S\sqrt{\frac{n-1}{\chi^2_{1-\alpha/2}}} \right)$. In the present problem, the confidence interval is

$$\left(\sqrt{0.597\left(\frac{14}{26.12}\right)}, \sqrt{0.597\left(\frac{14}{5.62}\right)} \right) = (0.5657, 1.2195).$$

(c) The center of the confidence interval in part (b) is $\frac{0.5657+1.2195}{2} = 0.8926$, which is not the same as $s = 0.597$.

9.43 The population distribution is assumed to be normal. We wish to test the hypotheses: $H_0: \sigma = 0.6$ versus $H_1: \sigma < 0.6$. The test statistic is $\chi^2 = \frac{(n-1)S^2}{\sigma^2} = \frac{(n-1)S^2}{(0.6)^2}$. For sample size $n = 40$ and $\alpha = 0.05$, we have $\text{d.f.} = n-1 = 39$ and $\chi^2_{0.95} \approx 25.7$. Since H_1 is left-sided, the rejection region is $R: \chi^2 \le \chi^2_{0.95} \approx 25.7$. From the sample data, we have $s = 0.475$. The value of the observed χ^2 is then $\chi^2 = \frac{39(0.475)^2}{(0.6)^2} = 24.4$, which lies in R. Hence, H_0 is rejected at $\alpha = 0.05$. As such, there is strong evidence that the red pine population standard deviation is smaller than 0.6.

9.45 The population distribution is assumed to be normal. A 95% confidence interval for σ is given by $\left(S\sqrt{\frac{n-1}{\chi^2_{0.025}}}, S\sqrt{\frac{n-1}{\chi^2_{0.975}}} \right)$, where d.f. = $n-1$. For sample size $n = 10$, we have $\text{d.f.} = n-1 = 9$ and $\chi^2_{0.025} = 19.02$ and $\chi^2_{0.975} = 2.70$. From the sample data, we have $s = 2.2706$. So, the 95% confidence interval for σ is given by

$$\left(S\sqrt{\frac{n-1}{\chi^2_{0.025}}}, S\sqrt{\frac{n-1}{\chi^2_{0.975}}} \right) = \left(2.2706\sqrt{\frac{9}{19.02}}, 2.2706\sqrt{\frac{9}{2.70}} \right) = (1.56, 4.15).$$

9.47 (a) A 90% confidence interval for σ is given by $\left(S\sqrt{\frac{n-1}{\chi^2_{0.05}}}, S\sqrt{\frac{n-1}{\chi^2_{0.95}}} \right)$, where d.f. = $n-1$. For the lizard length data, we have sample size $n = 20$, so that $\text{d.f.} = n-1 = 19$ and $\chi^2_{0.05} = 30.14$ and $\chi^2_{0.95} = 10.12$. From the sample data, we have $s = 20.143$. So, the 90% confidence interval for σ is given by

$$\left(S\sqrt{\frac{n-1}{\chi^2_{0.05}}}, S\sqrt{\frac{n-1}{\chi^2_{0.95}}} \right) = \left(20.143\sqrt{\frac{19}{30.14}}, 20.143\sqrt{\frac{19}{10.12}} \right) = (15.99, 27.60).$$

(b) The value $\sigma_0 = 9$ is not in the 90% confidence interval, so the null hypothesis $H_0 : \sigma = 9$ is rejected at $\alpha = 0.10$.

9.49 A 95% confidence interval for σ is given by $\left(S\sqrt{\frac{n-1}{\chi^2_{0.025}}}, S\sqrt{\frac{n-1}{\chi^2_{0.975}}} \right)$, where d.f. = $n-1$. For this data, we have sample size $n = 13$, so that d.f. $= n-1 = 12$ and $\chi^2_{0.025} = 23.34$ and $\chi^2_{0.975} = 4.40$. From the sample data, we have $s = 6.56$. So, the 95% confidence interval for σ is given by

$$\left(S\sqrt{\frac{n-1}{\chi^2_{0.025}}}, S\sqrt{\frac{n-1}{\chi^2_{0.975}}} \right) = \left(6.56\sqrt{\frac{12}{23.34}}, 6.56\sqrt{\frac{12}{4.40}} \right) = (4.70, 10.83).$$

9.51 (a) $t_{0.05} = 1.895$

(b) $t_{0.025} = 2.201$

(c) $-t_{0.05} = -1.895$

(d) $-t_{0.05} = -1.796$

9.53 (a) From the t-table we find that, with d.f. = 12, $t_{0.05} = 1.782$. Now, using the normal table, we interpolate to obtain

$$P[Z > t_{0.05}] = P[Z > 1.782] = 0.037.$$

Note that this probability is smaller than $P[T > t_{0.05}] = 0.05$.

(b) For d.f. = 5, we have $t_{0.05} = 2.015$. Using the normal table, we find that $P[Z > 2.015] = 0.022$, which is smaller than 0.05. Next, for d.f. = 20, we have $t_{0.05} = 1.725$, while the normal table gives $P[Z > 1.725] = 0.042$. We observe that the probability $P[Z > t_{0.05}]$ is always less than 0.05, but the difference decreases as the d.f. increases.

9.55 Assume that the measurements are normally distributed. A 99% confidence interval for the true mean μ is given by $\overline{X} \pm t_{0.005}\frac{S}{\sqrt{n}}$, where d.f. = $n-1$. For sample size $n = 14$, we have d.f. $= n-1 = 13$ and $t_{0.005} = 3.012$. From the sample data, we have $\overline{x} = 47$ and $s = 9.4$. The 99% confidence interval for the mean measurement μ is then given by

$$47 \pm 3.012\left(\frac{9.4}{\sqrt{14}}\right) = 47 \pm 7.6 \text{ or } (39.4, 54.6).$$

9.57 (a) The point estimate for μ is $\overline{x} = 3.007$.

The 95% error margin is $t_{0.025}\frac{S}{\sqrt{n}} = 2.145\left(\frac{0.5374}{\sqrt{15}}\right) = 0.298$

(b) A 90% confidence interval for the true mean μ is given by $\overline{X} \pm t_{0.05} \frac{S}{\sqrt{n}}$, where d.f. $= n-1$. For sample size $n=15$, we have d.f. $= n-1 = 14$ and $t_{0.05} = 1.761$. The 90% confidence interval for μ is then given by

$$3.007 \pm 2.145(0.139) = 3.007 \pm 0.298 \quad \text{or} \quad (2.709,\ 3.305).$$

(c) By the standard interpretation, 90% of all such confidence intervals would contain the true value of the mean.

9.59 Assume a normal population.

(a) We are to test the hypotheses $H_0 : \mu = 42$ versus $H_1 : \mu < 42$.

(b) The test statistic is $T = \frac{\overline{X} - 42}{S / \sqrt{21}}$.

(c) For sample size $n = 21$, we have d.f. $= n-1 = 20$ and $-t_{0.01} = -2.528$. Since H_1 is left-sided, the rejection region is $R : T \le -2.528$.

(d) From the sample data, we have $\overline{x} = 38.4$ and $s = 5.1$. The value of the observed t is then $t = \frac{38.4 - 42}{5.1 / \sqrt{21}} = -3.23$, which lies in R. Hence, H_0 is rejected at $\alpha = 0.01$. Furthermore, the associated p-value is $P[T \le -3.23] < 0.005$. So, there is strong evidence that the mean time to blossom is less than 42 days.

(e) Since H_0 is rejected, we could have made a Type I error, meaning that the mean time to blossom really is 42 days.

(f) A p-value <0.005 (from (d)) suggests very strong evidence that the mean blossom time is less than 42 days.

9.61 Since H_1 is two-sided, we need a two-sided rejection region. For sample size $n = 13$ and $\alpha = 0.02$, we have d.f. $= n-1 = 12$ and $t_{0.01} = 2.681$. So, the rejection region in this case is $R : |T| \ge 2.681$. Since the observed value of t, namely 2.736, lies in R, H_0 is rejected at $\alpha = 0.02$.

9.63 (a) Let μ denote the mean potency after exposure. The supplier's claim is valid if $\mu > 65$, and this is to be demonstrated. So, we are to test the hypotheses $H_0 : \mu = 65$ versus $H_1 : \mu > 65$ with $\alpha = 0.05$.

(b) The test statistic is $T = \frac{\overline{X} - 65}{S / \sqrt{9}}$. For sample size $n = 9$, we have d.f. $= n-1 = 8$ and $t_{0.05} = 1.860$. Since H_1 is right-sided, the rejection region is $R : T \ge 1.860$. Assume a normal population.

(c) From the sample data, we have $\overline{x} = 65.22$ and $s = 3.67$. The value of the observed t is then $t = \frac{65.22 - 65}{3.67 / \sqrt{9}} = 0.18$, which does not lie in R. Hence, H_0 is not rejected at $\alpha = 0.05$. In fact, a comparison of the rejection region (left boundary of 1.860) and this value of the *t*-statistic (0.18) clearly shows that the null hypotheses would not be rejected at any reasonable level of α. As such, the claim that $\mu > 65$ is not demonstrated.

9.65 Assume a normal population. We want to establish that $\mu > 1500$ (i.e., the advertiser's claim is false). So, we are to test the hypotheses $H_0 : \mu = 1500$ versus $H_1 : \mu > 1500$ with $\alpha = 0.05$. The test statistic is $T = \dfrac{\overline{X} - 1500}{S/\sqrt{5}}$. For sample size $n = 5$, we have $\text{d.f.} = n - 1 = 4$ and $t_{0.05} = 2.132$. Since H_1 is right-sided, the rejection region is $R : T \geq 2.132$. From the sample data, we have $\overline{x} = 1620$ and $s = 90$. The value of the observed *t* is then $t = \dfrac{1620 - 1500}{90/\sqrt{5}} = 2.98$, which lies in *R*. Hence, H_0 is rejected at $\alpha = 0.05$. Furthermore, scanning the *t*-table for d.f. = 4, we find that 2.98 lies between $t_{0.025} = 2.776$ and $t_{0.01} = 3.747$. So, the associated *p*-value is $P[T \geq 2.98]$ is between 0.01 and 0.025, near about 0.02. As such, we conclude that the advertiser's claim is strongly contradicted.

9.67 We use Minitab for this exercise. One could also complete it by hand by mimicking the approach of Exercise 9.27, for instance.

(a) Enter the sample data into column C1 of a Minitab worksheet. The output is as follows:

T Confidence Intervals

Variable	N	Mean	StDev	SE Mean	95.0 % CI
C1	7	439.57	17.82	6.74	(423.08, 456.06)

So, the 95% confidence interval is (423.08, 456.06).

(b) In order to test the hypotheses $H_0 : \mu = 453$ versus $H_1 : \mu \neq 453$, run a one-sample *t* test on Minitab (using the data in column C1) to obtain:

T-Test of the Mean

Test of mu = 453.00 vs mu not = 453.00

Variable	N	Mean	StDev	SE Mean	T	P
C1	7	439.57	17.82	6.74	-1.99	0.093

Hence, since the *p*-value is 0.093, we (barely) reject $_0$ at $\alpha = 0.10$.

9.69(a) $\chi^2_{0.05} = 12.59$

(b) $\chi^2_{0.025} = 36.78$

(c) $\chi^2_{0.95} = 1.64$

(d) $\chi^2_{0.975} = 10.98$

9.71 Assume a normal population.

(a) We are to test the hypotheses $H_0 : \mu = 55.0$ versus $H_1 : \mu \neq 55.0$.

(b) The test statistic is $T = \dfrac{\overline{X} - 55.0}{S / \sqrt{11}}$.

(c) For sample size $n = 11$, we have $\text{d.f.} = n - 1 = 10$ and $t_{0.025} = 2.228$. Since H_1 is two-sided, the rejection region is $R : |T| \geq 2.228$.

(d) From the sample data (computed in Exercise 9.68), the value of the observed t is then $t = \dfrac{58.7 - 55.0}{4.45 / \sqrt{11}} = 2.758$, which lies in R. Hence, H_0 is rejected at $\alpha = 0.05$.

(e) Since H_0 is rejected, we could have made a Type I error, meaning that the mean time to finish a bottle really is 42 days.

(f) The p-value is $2\,P(T \geq 2.758) = 0.02$, which is strong evidence in favor of the mean life being different from 55 days.

9.73 The population distribution is assumed to be normal.
Let σ denote the standard deviation in bacteria count per unit volume of water.
We wish to test the hypotheses: $H_0 : \sigma = 18.0$ versus $H_1 : \sigma < 18.0$.

The test statistic is $\chi^2 = \dfrac{(n-1)S^2}{\sigma^2} = \dfrac{(n-1)S^2}{18^2}$. For sample size $n = 10$, we have $\text{d.f.} = n - 1 = 9$. From the sample data, we have $s = 10.81$ The value of the observed χ^2 is then $\chi^2 = \dfrac{9(10.81)^2}{18^2} = 3.246$. The associated p-value is $P\left(\chi^2 \leq 3.246\right) < 0.05$, so that there is strong evidence in support of the claim that $\sigma < 18.0$.

9.75 (a) Uncertain since the true mean length (a <u>population</u> parameter) is not known. Refer to the text for a detailed explanation.

(b) The very definition of a confidence interval ensures this statement is true – see property 2 on page 302 of the text.

9.77 Use only those values that are coded with a 1 (for Male). Enter into column C2 of a Minitab worksheet – here is the output.

T-Test of the Mean

```
Test of mu = 2.700 vs mu > 2.700

Variable     N      Mean     StDev     SE Mean       T          P
  C2        20     2.964     0.525       0.117     2.25      0.018
```

Hence, we reject H_0 at $\alpha = 0.025$ since the p-value is less than 0.025.

Chapter 10
COMPARING TWO TREATMENTS

10.1 First group using first letter:

{B, C}	{B, E}	{B, H}	{B, P}
{C, E}	{C, H}	{C, P}	
{E, H}	{E, P}		
{H, P}			

10.3 (a) We use the first letter of the first names:

{ (S, G), (T, E) }	{ (S, G), (T, R) }	{ (S, G), (E, R) }
{ (S, J), (T, E) }	{ (S, J), (T, R) }	{ (S, J), (E, R) }
{ (J, G), (T, E) }	{ (J, G), (T, R) }	{ (J, G), (E, R) }

(b) There are three sets each consisting of three pairs.

10.5 (a) (i) Yes, distinct observations from within a single random sample are assured to be independent.

(ii) Yes, the elements of a sample are independent and identically distributed.

(b) (i) Yes, the two samples are assumed to be independent.
(ii) Not necessarily. The two different samples are pulled from different populations and hence, are independent.

10.7 (a) A point estimate of $\mu_1 - \mu_2$ is given by $\bar{x} - \bar{y} = 73 - 66 = 7$.

$$\text{Estimated S.E.} = \sqrt{\frac{s_1^2}{n_1} + \frac{s_2^2}{n_2}} = \sqrt{\frac{151}{52} + \frac{142}{44}} = 2.48$$

(b) A large sample 95% confidence interval for $\mu_1 - \mu_2$ is given by

$$(\bar{x}-\bar{y}) \pm z_{0.05/2}\sqrt{\frac{s_1^2}{n_1}+\frac{s_2^2}{n_2}} = 7 \pm 1.96(2.48) = 7 \pm 4.9 \text{ or } (2.1, 11.9).$$

10.9 We are to test the hypotheses $H_0: \mu_1 - \mu_2 = 0$ versus $H_1: \mu_1 - \mu_2 \neq 0$ with $\alpha = 0.05$. Since the sample sizes are large, we employ the *Z*-test, and so, the test statistic is $Z = \dfrac{\overline{X}-\overline{Y}}{\sqrt{\dfrac{S_1^2}{n_1}+\dfrac{S_2^2}{n_2}}}$. Since H_1 is two-sided, the rejection region is $R: |Z| \geq z_{0.05/2} = 1.96$. From the sample data, we calculate the value of the observed *z* to be $z = \dfrac{76.4-81.2}{\sqrt{\dfrac{(8.2)^2}{90}+\dfrac{(7.6)^2}{100}}} = -\dfrac{4.8}{1.151} = -4.17$, which lies in *R*. Hence, H_0 is rejected at $\alpha = 0.05$. Furthermore, the associated *p*-value is $P[|Z| \geq 4.17] < 0.0001$, so the evidence against *no mean difference* is very strong.

10.11(a) Since the assertion is that $\mu_1 > \mu_2$, we formulate the hypotheses $H_0: \mu_1 - \mu_2 = 2$ versus $H_1: \mu_1 - \mu_2 > 2$.

(b) Since the sample sizes $n_1 = 50$ and $n_2 = 50$ are large, we employ the *Z*-test, and so, the test statistic is $Z = \dfrac{\overline{X}-\overline{Y}}{\sqrt{\dfrac{S_1^2}{n_1}+\dfrac{S_2^2}{n_2}}}$. Since H_1 is right-sided and $\alpha = 0.05$, the rejection region is $R: Z \geq z_{0.05} = -1.645$.

(c) From the sample data, we calculate the value of the observed *z* to be $z = \dfrac{12.6-9.5}{\sqrt{\dfrac{(4.2)^2}{50}+\dfrac{(1.9)^2}{50}}} = \dfrac{3.1}{0.652} = 1.69$, which lies in *R*. Hence, H_0 is rejected at $\alpha = 0.05$. Furthermore, the associated *p*-value is $P[Z \geq 4.755] = 0.0455$, so the evidence is support of H_1 is very strong.

10.13 The problem here is to test $H_0: \mu_1 - \mu_2 = 0$ versus $H_1: \mu_1 - \mu_2 \neq 0$ with $\alpha = 0.05$. Because 0 is not included in the 95% confidence interval (0.0993, 0.3867) (from Exercise 10.10), the conclusion is that H_0 is rejected at $\alpha = 0.05$.

10.15 Let μ_W = mean of walnut-eating group and μ_C = mean of control group. The question is whether or not $\mu_W > \mu_C$. So, we formulate the hypotheses:

$$H_0: \mu_W - \mu_C = 0 \text{ versus } H_1: \mu_W - \mu_C > 0$$

The sample sizes are $n_W = 55$ and $n_C = 52$ are large so we use the Z-test. The test statistic is $Z = \dfrac{\overline{X} - \overline{Y}}{\sqrt{\dfrac{S_W^2}{n_W} + \dfrac{S_C^2}{n_C}}}$. Use the sample data yields $Z = 2.94$. The associated p-value is $P(Z > 2.94) \approx 0.0016$, so we conclude that μ_W is significantly higher than μ_C.

10.17 The sample sizes are large so we use the Z-test. The test statistic is $Z = \dfrac{\overline{X} - \overline{Y}}{\sqrt{\dfrac{S_A^2}{n_A} + \dfrac{S_{Not-A}^2}{n_{Not-A}}}} = \dfrac{2.48 - 1.57}{\sqrt{\dfrac{1.94^2}{52} + \dfrac{1.31^2}{67}}} \approx 2.907$. The associated p-value is $P(Z > 2.907) \approx 0.003$, so we conclude that μ_A is significantly higher than μ_{Non-A}.

10.19 (a) We first obtain:

$\overline{x} = 5$ $\qquad$ $\overline{y} = 7$

$$s_1^2 = \frac{\left(1^2 + (-1)^2 + 2^2 + (-2)^2\right)}{3} = 3.333 \qquad s_2^2 = \frac{\left(1^2 + 2^2 + (-1)^2\right)}{2} = 3$$

Consequently, the pooled variance is given by

$$s_{\text{pooled}}^2 = \frac{s_1^2(n_1 - 1) + s_2^2(n_2 - 1)}{n_1 + n_2 - 2} = \frac{10 + 6}{4 + 3 - 2} = 3.2.$$

(b) We estimate the common sigma by $s_{\text{pooled}} = \sqrt{3.2} = 1.789$.

(c) The t-statistic is $T = \dfrac{\overline{X} - \overline{Y}}{s_{\text{pooled}}\sqrt{\dfrac{1}{n_1} + \dfrac{1}{n_2}}}$ with d.f. $= n_1 + n_2 - 2$. In the present problem, we have $t = \dfrac{5 - 7}{1.79\sqrt{\dfrac{1}{4} + \dfrac{1}{3}}} = -1.463$, with d.f. = 5.

10.21 (a) $s_{\text{pooled}}^2 = \dfrac{\sum (x_i - \overline{x})^2 + \sum (y_i - \overline{y})^2}{n_1 + n_2 - 2} = \dfrac{28 + 32}{14 + 13 - 2} = 2.4$

(b) We are to test the hypotheses $H_0 : \mu_1 - \mu_2 = 0$ versus $H_1 : \mu_1 - \mu_2 > 0$ with $\alpha = 0.05$. We assume normal populations with equal variance, so the test statistic is

$$T = \frac{\overline{X} - \overline{Y}}{s_{\text{pooled}}\sqrt{\dfrac{1}{n_1} + \dfrac{1}{n_2}}} \quad \text{with d.f.} = n_1 + n_2 - 2$$

Since H_1 is right-sided, the rejection region is $R: T \geq t_{0.05} = 1.708$. From the sample data, we calculate the value of the observed t is

$$t = \frac{9-17}{\sqrt{2.4}\sqrt{\frac{1}{14}+\frac{1}{13}}} = \frac{-8}{0.597} = -13.40,$$

which does not lie in R. So, H_0 is not rejected at $\alpha = 0.05$.

(c) A corresponding 95% confidence interval has the form

$$\left(\overline{X}-\overline{Y}\right) \pm t_{\alpha/2}\, s_{\text{pooled}} \sqrt{\frac{1}{n_1}+\frac{1}{n_2}}\ .$$

Since $t_{0.025} = 2.060$ for d.f. = 25, the 95% confidence interval in this case is $-8 \pm 2.060(0.597) = -8 \pm 1.230$ or (-9.23, -7.23).

10.23 (a) The summary statistics are:

Method 1:	$n_1 = 10$,	$\overline{x} = 19.1$,	$s_1 = 4.818$
Method 2:	$n_2 = 10$,	$\overline{y} = 23.3$,	$s_2 = 5.559$

Denote by μ_1 and μ_2 the population mean job times corresponding to Method 1 and Method 2, respectively. Since the conjecture is that μ_1 is smaller than μ_2, we formulate the hypotheses

$$H_0: \mu_1 - \mu_2 = 0 \text{ versus } H_1: \mu_1 - \mu_2 < 0\ .$$

We assume normal populations with equal variance, so the test statistic is

$$T = \frac{\overline{X}-\overline{Y}}{s_{\text{pooled}}\sqrt{\frac{1}{n_1}+\frac{1}{n_2}}} \quad \text{with d.f.} = n_1 + n_2 - 2$$

Since H_1 is left-sided, the rejection region is $R: T \leq -t_{0.05} = -1.734$ for d.f. = 10 + 10 – 2 = 18. Using the summary statistics, we obtain

$$\overline{x} - \overline{y} = 19.1 - 23.3 = -4.2$$

$$s_{\text{pooled}} = \sqrt{\frac{(n_1-1)s_1^2 + (n_2-1)s_2^2}{n_1+n_2-2}} = \sqrt{\frac{9(4.818)^2 + 9(5.559)^2}{18}} = 5.201$$

Hence, the observed value of t is given by

$$t = \frac{-4.2}{5.201\sqrt{\frac{1}{10}+\frac{1}{10}}} = -1.81,$$

which lies in R. So, H_0 is rejected at $\alpha = 0.05$. We conclude that the mean job time is significantly less for Method 1 than for Method 2.

(b) See part (a).

(c) A corresponding 95% confidence interval has the form

$$\left(\overline{X}-\overline{Y}\right)\pm t_{0.05/2}\, s_{\text{pooled}}\sqrt{\frac{1}{n_1}+\frac{1}{n_2}}\,.$$

Since $t_{0.025}=2.101$ for d.f. = 18, using the calculations in part (a), we obtain that the 95% confidence interval in this case is

$-4.2\pm 2.101(2.326)=-4.2\pm 4.89$ or $(-9.09, 0.69)$.

10.25 (a) Use the *Z*-test since the sample sizes are large.

(b) Use the *t*-test with pooling. The sample sizes are small, and s_1 and s_2 are not too far apart ($s_2/s_1=1.3$). Assume normal populations with equal variances.

(c) Use the conservative *t*-test without pooling. The sample sizes are small, and s_2/s_1 is larger than 2. Assume that the populations are normal.

(d) Use the *Z*-test since the sample sizes are large.

10.27 (a) Diet, sleep, and additional exercise can also impact the amount of reduction in this 5-week period.

(b) Try to control for the above variables when breaking them into groups. Identifying a sample where the diet, sleep, etc was fairly similar among everyone in the sample is important; then, assign 10 randomly to each exercise method.

10.29 (a) The summary statistics are:

$$n_1=9,\qquad \overline{x}=1.887,\qquad s_1^2=0.0269$$
$$n_2=9,\qquad \overline{y}=0.670,\qquad s_2^2=0.1133$$

The more conservative test in Section 10.5.1 requires that we use $t_{\alpha/2}$ with d.f. = 8 and $\alpha=0.05$, which is $t_{0.025}=2.306$. So, a 95% confidence interval for $\mu_1-\mu_2$ has the form

$$\left(\overline{X}-\overline{Y}\right)\pm t_{0.025}\sqrt{\frac{S_1^2}{n_1}+\frac{S_2^2}{n_2}}=1.217\pm 2.306(0.1247)$$

or $(0.9295, 1.5045)$.

(b) The confidence interval from Example 12 is 0.94, 1.49). The confidence interval in (a) is larger, which is reasonable since it is more conservative.

10.31 We drew slips with α, β, and τ, so group 1 is {alpha, beta, tau}. (Answers may vary.)

10.33 We must be careful. It is likely that mothers who are warmer toward everyone have a much higher tendency to nurse their babies than mothers who have colder personalities. There are many other reasons as well.

10.35 (a) The $n = 6$ paired differences $d = x - y$ are 3, 3, 5, 1, 2, -2.
Their mean and standard deviation are $\bar{d} = 2$, $s_D = 2.366$.

So, $t = \dfrac{2}{2.366/\sqrt{6}} = 2.071$.

(b) d.f. = $n - 1 = 5$

10.37 It is a matched pair sample because there may be considerable variation of conditions in the different plants. The paired differences d = (before – after) are $2, 1, -1, 2, 3, -1$. We assume these differences constitute a random sample from a normal distribution with mean δ. The null hypothesis of no change is $H_0 : \delta = 0$, and the alternative of more loss before than after is $H_1 : \delta > 0$. We calculate

$$\bar{d} = 1.0,\ \ s_D = 1.673,\ \ \tfrac{s_D}{\sqrt{6}} = 0.683,\ \ t = \tfrac{1.0}{0.683} = 1.46, \text{ and d.f.} = 5.$$

With $\alpha = 0.05$, the rejection region is $R : T > t_{0.05} = 2.015$. Since the observed t is less than 2.015, H_0 is not rejected at $\alpha = 0.05$. As such, the claim of effectiveness of the safety program is not demonstrated.

10.39 This is a matched pair design because from each farm a pair of milk specimens are taken, and then one is treated with PC while the other is not.

(a) The paired differences d = (with PC – without PC) are $7, 6, -2, 10, 0, 8, 4$.
From this, we obtain the following summary statistics:

$$n = 7,\ \ \bar{d} = 4.714, \text{ and } s_D = 4.348$$

We assume that the population distribution of the D's is normal, and denote the population mean by δ. Because the conjecture is that the mean response with PC is higher than without PC (that is, $\delta > 0$), we formulate the hypotheses

$$H_0 : \delta = 0 \ \text{ versus } \ H_1 : \delta > 0.$$

The test statistic is $T = \dfrac{\bar{D}}{S_D/\sqrt{n}}$, d.f. = $n - 1$. Since H_1 is right-sided and $\alpha = 0.05$, the rejection region is $R : T \geq t_{0.05} = 1.943$ (for d.f. = 6). Using the summary statistics, we see that the observed t is $t = \frac{4.714}{4.348/\sqrt{7}} = 2.87$, which lies in R. Hence, H_0 is rejected at $\alpha = 0.05$. Furthermore, the associated p-value is $P[T \geq 2.87] = 0.0142$. As such, we conclude that there is strong evidence in support of this conjecture.

(b) The corresponding 90% confidence interval for δ is given by

$$\bar{d} \pm t_{0.05}\tfrac{s_D}{\sqrt{n}} = 4.71 \pm 1.943\left(\tfrac{4.35}{\sqrt{7}}\right) = 4.71 \pm 3.19 \ \text{ or } \ (1.52, 7.90).$$

10.41 (a) We formulate the hypotheses

$$H_0: \delta = 0 \text{ versus } H_1: \delta \neq 0.$$

The test statistic is $T = \dfrac{\overline{D}}{S_D / \sqrt{n}}$, d.f. = $n-1$. Since H_1 is two-sided and $\alpha = 0.05$, the rejection region is $R: |T| \geq t_{0.025} = 2.646$ (for d.f. = 22). Using the summary statistics, we see that the observed t is $t = \frac{-8.0}{14.5/\sqrt{23}} = -2.646$, which lies just barely in R. Hence, H_0 is rejected at $\alpha = 0.05$. This suggests that the data provides strong enough evidence that the mean difference is not zero.

(b) The corresponding 95% confidence interval for δ is given by $\bar{d} \pm t_{0.025} \frac{s_D}{\sqrt{n}} = -8.0 \pm 2.080\left(\frac{14.5}{\sqrt{23}}\right) = -8.0 \pm 6.289$ or (-14.289, -1.711).

10.43 This is a matched pair design. The paired differences d = (left – right) are given, from which we obtain the following summary statistics:

$$n = 10,\ \bar{d} = 60.37, \text{ and } s_D = 0.1418$$

We assume that the population distribution of these differences is normal, and denote the population mean by δ.

(a) We formulate the hypotheses

$$H_0: \delta = 60 \text{ versus } H_1: \delta \neq 60.$$

The test statistic is $T = \dfrac{\overline{D} - \delta}{S_D / \sqrt{n}}$, d.f. = $n-1$. Since H_1 is two-sided and $\alpha = 0.05$, the rejection region is $R: |T| \geq t_{0.025} = 2.262$ (for d.f. = 9). Using the summary statistics, we see that the observed t is $t = \frac{60.37-60}{0.1418/\sqrt{10}} = 8.251$, which lies in R. Hence, H_0 is rejected at $\alpha = 0.05$.

(b) For $\alpha = 0.05$ and d.f. = 9, we have $t_{0.025} = 2.262$. So, the 95% confidence interval for δ is given by

$$\bar{d} \pm t_{0.025} \tfrac{s_D}{\sqrt{n}} = 60.37 \pm 2.262(0.045) = 60.37 \pm 0.10179$$

or (60.268, 60.472).

10.45 (a) From the data in Exercise 10.10, we see that

$$\hat{p}_1 = \tfrac{180}{226} = 0.796,\ \hat{p}_2 = \tfrac{149}{247} = 0.603, \text{ so that}$$

$$\hat{p}_1 - \hat{p}_2 = 0.796 - 0.603 = 0.193$$

Estimated S.E. $= \sqrt{\dfrac{\hat{p}_1\hat{q}_1}{n_1} + \dfrac{\hat{p}_2\hat{q}_2}{n_2}} = \sqrt{\dfrac{(0.796)(0.204)}{226} + \dfrac{(0.603)(0.397)}{247}} = 0.0411$

So, a 95% confidence interval for $p_1 - p_2$ is given by

$$(\hat{p}_1 - \hat{p}_2) \pm z_{0.05/2}\sqrt{\frac{\hat{p}_1\hat{q}_1}{n_1} + \frac{\hat{p}_2\hat{q}_2}{n_2}} = 1.93 \pm 1.96(0.0411) = 1.93 \pm 0.0806$$

or (1.85, 2.01).

(b) For such a test, the test statistic is $Z = \dfrac{\hat{p}_1 - \hat{p}_2}{\sqrt{\hat{p}\hat{q}}\sqrt{\dfrac{1}{n_1}+\dfrac{1}{n_2}}}$.

Observe that

Pooled estimate $\hat{p} = \dfrac{n_1\hat{p}_1 + n_2\hat{p}_2}{n_1+n_2} = \dfrac{226(0.796)+247(0.603)}{226+247} = 0.381$,

Observed z is $\dfrac{1.93}{\sqrt{(0.619)(0.381)}\sqrt{\dfrac{1}{226}+\dfrac{1}{247}}} = \dfrac{1.93}{0.0217} = 88.94$.

The associated p-value is $P[Z \geq 85.99] \approx 0$. This means that H_0 would be rejected for small values of $\alpha < 0.05$.

10.47 Let p_1 and p_2 denote the probabilities of 'resistant' for the HRL and LRL groups, respectively. We formulate the hypotheses:

$$H_0 : p_1 = p_2 \text{ versus } H_1 : p_1 < p_2.$$

The test statistic is $Z = \dfrac{\hat{p}_1 - \hat{p}_2}{\sqrt{\hat{p}\hat{q}}\sqrt{\dfrac{1}{n_1}+\dfrac{1}{n_2}}}$. Since H_1 is left-sided, the rejection region is of the form $R: Z \leq c$. We calculate the following:

$\hat{p}_1 = \frac{15}{49} = 0.306, \quad \hat{p}_2 = \frac{42}{54} = 0.778$

Pooled estimate $\hat{p} = \dfrac{n_1\hat{p}_1 + n_2\hat{p}_2}{n_1+n_2} = \dfrac{15+42}{49+54} = 0.553$,

Observed z is $\dfrac{0.306-0.778}{\sqrt{(0.553)(0.447)}\sqrt{\dfrac{1}{49}+\dfrac{1}{54}}} = -4.81$.

The associated p-value is $P[Z \leq -4.81]$ is less than 0.0001. So, there is very strong evidence in support of H_1.

10.49 Let p_1 and p_2 denote the probability of survival for the treated group (with carbolic acid) and the control group (without carbolic acid), respectively. We formulate the hypotheses:

$$H_0 : p_1 = p_2 \text{ versus } H_1 : p_1 \neq p_2.$$

The test statistic is $Z = \dfrac{\hat{p}_1 - \hat{p}_2}{\sqrt{\hat{p}\hat{q}}\sqrt{\dfrac{1}{n_1}+\dfrac{1}{n_2}}}$. Since H_1 is two-sided and $\alpha = 0.05$, the rejection region is $R: |Z| \geq z_{0.05/2} = 1.96$.

We calculate the following:

$\hat{p}_1 = \frac{34}{40} = 0.850, \quad \hat{p}_2 = \frac{19}{35} = 0.543$

Pooled estimate $\hat{p} = \frac{n_1\hat{p}_1 + n_2\hat{p}_2}{n_1 + n_2} = \frac{34+19}{40+35} = 0.707$,

Observed z is $\frac{0.850 - 0.543}{\sqrt{(0.707)(0.293)}\sqrt{\frac{1}{40}+\frac{1}{35}}} = 2.91$,

which lies in R. Hence, H_0 is rejected at $\alpha = 0.05$. Furthermore, the associated p-value $2P[Z \le -2.91] = 2(0.0018) = 0.0036$. This means that H_0 would be rejected with α as small as 0.0036. As such, a difference in the survival rates is strongly demonstrated by the data.

10.51 Let p_1 and p_2 denote the population proportions of ≤ 8 hours of sleep for the age group 30 – 40, and the age group 60 – 70, respectively. We formulate the hypotheses: $H_0 : p_1 = p_2$ versus $H_1 : p_1 > p_2$.

The test statistic is $Z = \frac{\hat{p}_1 - \hat{p}_2}{\sqrt{\hat{p}\hat{q}}\sqrt{\frac{1}{n_1}+\frac{1}{n_2}}}$. We calculate the following:

$\hat{p}_1 = \frac{173}{250} = 0.692, \quad \hat{p}_2 = \frac{120}{250} = 0.480$

Pooled estimate $\hat{p} = \frac{n_1\hat{p}_1 + n_2\hat{p}_2}{n_1 + n_2} = \frac{293}{500} = 0.586$,

Observed z is $\frac{0.692 - 0.480}{\sqrt{(0.586)(0.414)}\sqrt{\frac{1}{250}+\frac{1}{250}}} = 4.81$

The associated p-value is $P[Z \ge 4.81]$ is less than 0.0002. So, there is very strong evidence in support of H_1.

10.53 (a) Let p_1 and p_2 denote the probability of having prominent wrinkles for smokers and non-smokers, respectively. We formulate the hypotheses:

$$H_0 : p_1 = p_2 \text{ versus } H_1 : p_1 > p_2.$$

The test statistic is $Z = \frac{\hat{p}_1 - \hat{p}_2}{\sqrt{\hat{p}\hat{q}}\sqrt{\frac{1}{n_1}+\frac{1}{n_2}}}$. We calculate the following:

$\hat{p}_1 = \frac{95}{150} = 0.633, \quad \hat{p}_2 = \frac{103}{250} = 0.412$

Pooled estimate $\hat{p} = \frac{n_1\hat{p}_1 + n_2\hat{p}_2}{n_1 + n_2} = \frac{95+103}{400} = 0.495$,

Observed z is $\frac{0.633 - 0.412}{\sqrt{(0.495)(0.505)}\sqrt{\frac{1}{150}+\frac{1}{250}}} = 4.28$

The associated p-value is $P[Z \ge 4.28]$ is less than 0.0002. So, there is very strong evidence in support of H_1.

(b) A direct causal relation between smoking and wrinkled skin cannot be readily concluded. Various psycho-physiological factors could influence both the smoking habit and the presence of wrinkled skin.

10.55 (a) Let p_1 and p_2 denote the probability of getting hepatitis for the 'vaccine' group and the 'placebo' group, respectively. We formulate the hypotheses:

$$H_0: p_1 = p_2 \text{ versus } H_1: p_1 < p_2.$$

The test statistic is $Z = \dfrac{\hat{p}_1 - \hat{p}_2}{\sqrt{\hat{p}\hat{q}}\sqrt{\dfrac{1}{n_1}+\dfrac{1}{n_2}}}$. Since H_1 is left-sided and $\alpha = 0.01$, the rejection region is $R: Z \le -z_{0.01} = -2.33$. We calculate the following:

$$\hat{p}_1 = \tfrac{11}{549} = 0.020, \quad \hat{p}_2 = \tfrac{70}{534} = 0.131$$

Pooled estimate $\hat{p} = \dfrac{n_1\hat{p}_1 + n_2\hat{p}_2}{n_1 + n_2} = \dfrac{11+70}{1083} = 0.075$,

Observed z is $\dfrac{0.020 - 0.131}{\sqrt{(0.075)(0.925)}\sqrt{\dfrac{1}{549}+\dfrac{1}{534}}} = -6.93$,

which lies in R. Hence, H_0 is rejected at $\alpha = 0.01$. So, there is very strong evidence that the vaccine is effective.

(b) $\hat{p}_1 - \hat{p}_2 = 0.131 - 0.020 = 0.111$

Estimated S.E.= $\sqrt{\dfrac{\hat{p}_1\hat{q}_1}{n_1}+\dfrac{\hat{p}_2\hat{q}_2}{n_2}} = \sqrt{\dfrac{(0.020)(0.980)}{549}+\dfrac{(0.131)(0.869)}{534}}$

$= 0.016$

So, a 95% confidence interval for $p_1 - p_2$ is given by

$$(\hat{p}_1 - \hat{p}_2) \pm z_{0.05/2}\sqrt{\frac{\hat{p}_1\hat{q}_1}{n_1}+\frac{\hat{p}_2\hat{q}_2}{n_2}} = 0.111 \pm 1.96(0.016) = 0.111 \pm 0.031$$

or (0.08, 0.14).

is less than 0.0002. Hence, the data strongly substantiate H_1.

10.57 (a) $\hat{p}_E = \tfrac{78}{150} = 0.520, \quad \hat{p}_H = \tfrac{39}{160} = 0.244$

$\hat{p}_E - \hat{p}_H = 0.276$

Estimated S.E.= $\sqrt{\dfrac{\hat{p}_1\hat{q}_1}{n_1}+\dfrac{\hat{p}_2\hat{q}_2}{n_2}} = \sqrt{\dfrac{(0.520)(0.480)}{150}+\dfrac{(0.244)(0.756)}{160}}$

$= 0.053$

So, a 95% confidence interval for $p_E - p_H$ is given by

$$(\hat{p}_E - \hat{p}_H) \pm z_{0.05/2}\sqrt{\frac{\hat{p}_E\hat{q}_E}{n_1}+\frac{\hat{p}_H\hat{q}_H}{n_2}} = 0.276 \pm 1.96(0.053) = 0.276 \pm 0.104$$

or (0.17, 0.38).

(b) $\hat{p}_C = \frac{43}{200} = 0.215$, $\hat{p}_H = \frac{39}{160} = 0.244$, $\hat{p}_H - \hat{p}_C = 0.029$

Estimated S.E.= $\sqrt{\frac{\hat{p}_H\hat{q}_H}{n_1} + \frac{\hat{p}_C\hat{q}_C}{n_2}} = \sqrt{\frac{(0.244)(0.756)}{160} + \frac{(0.215)(0.785)}{200}}$

$= 0.045$

So, a 90% confidence interval for $p_H - p_C$ is given by

$$(\hat{p}_H - \hat{p}_C) \pm z_{0.10/2}\sqrt{\frac{\hat{p}_H\hat{q}_H}{n_1} + \frac{\hat{p}_C\hat{q}_C}{n_2}} = 0.029 \pm 1.645(0.045) = 0.029 \pm 0.074$$

or (–0.045, 0.103).

(c) A 95% confidence interval for a population proportion is given by

$\hat{p} \pm 1.96\sqrt{\frac{\hat{p}\hat{q}}{n}}$. Calculations for the individual groups are as follows:

Diabetes: $n = 160$, $\hat{p}_D = \frac{41}{160} = 0.256$

The confidence interval for p_D is given by:

$$0.256 \pm 1.96\sqrt{\frac{(0.256)(0.744)}{160}} = 0.256 \pm 0.068 \quad \text{or} \quad (0.19, 0.32)$$

Heart condition: $n = 160$, $\hat{p}_H = \frac{39}{160} = 0.244$

The confidence interval for p_H is given by:

$$0.244 \pm 1.96\sqrt{\frac{(0.244)(0.756)}{160}} = 0.244 \pm 0.067 \quad \text{or} \quad (0.18, 0.31)$$

Epilepsy: $n = 150$, $\hat{p}_E = \frac{78}{150} = 0.520$

The confidence interval for p_E is given by:

$$0.520 \pm 1.96\sqrt{\frac{(0.520)(0.480)}{150}} = 0.520 \pm 0.080 \quad \text{or} \quad (0.44, 0.60)$$

Control: $n = 20$, $\hat{p}_C = \frac{43}{200} = 0.215$

The confidence interval for p_C is given by:

$$0.215 \pm 1.96\sqrt{\frac{(0.215)(0.785)}{200}} = 0.215 \pm 0.057 \quad \text{or} \quad (0.16, 0.27)$$

10.59 (a) A large sample 95% confidence interval for $\mu_1 - \mu_2$ is given by

$$(\bar{x} - \bar{y}) \pm z_{0.05/2}\sqrt{\frac{s_1^2}{n_1} + \frac{s_2^2}{n_2}} = (12.2 - 7.2) \pm 1.96\sqrt{\frac{(1.1)^2}{40} + \frac{(3.4)^2}{40}}$$

$$= 5 \pm 1.96(0.565) = 5 \pm 1.1074$$

or (3.893, 6.107).

(b) You need to make certain that the group of 80 is completely randomized, in some way. Then, choose 40 at random from it to minimize any potential bias.

10.61 (a) Since the assertion is that $\mu_2 > \mu_1 + 150$, we formulate the hypotheses $H_0 : \mu_2 - \mu_1 = 150$ versus $H_1 : \mu_1 - \mu_2 > 150$.

(b) Since the sample sizes $n_1 = 18$ and $n_2 = 48$ are sufficiently large, we employ the *Z*-test, and so, the test statistic is $Z = \frac{(\overline{Y} - \overline{X}) - 150}{\sqrt{\frac{S_1^2}{n_1} + \frac{S_2^2}{n_2}}}$. Since H_1 is right-sided and $\alpha = 0.05$, the rejection region is $R : Z \geq z_{0.05} = 1.645$.

(c) From the sample data, we calculate the value of the observed *z* to be $z = \frac{(563.9 - 177.2) - 150}{132.156} = 1.829$, which lies in *R*. Hence, H_0 is rejected at $\alpha = 0.05$. Furthermore, the associated *p*-value is $P[Z \geq 1.829] = 0.0335$, so the evidence is support of H_1 is strong.

10.63 (a) A large sample 90% confidence interval for $\mu_A - \mu_B$ is given by

$$(\overline{x} - \overline{y}) \pm z_{0.1\%_2}\sqrt{\frac{s_1^2}{n_1} + \frac{s_2^2}{n_2}} = (4.64 - 4.03) \pm 1.645\sqrt{\frac{(1.25)^2}{55} + \frac{(1.82)^2}{58}}$$
$$= 0.61 \pm 0.48$$

or (0.13, 1.09). We are 90% confident that μ_A is 0.13 to 1.09 hours longer than μ_B.

(b) The 95% confidence interval for μ_A is given by

$$\overline{x} \pm z_{0.05\!/_2}\frac{s_1}{\sqrt{n_1}} = 4.64 \pm 1.96\left(\frac{1.25}{\sqrt{55}}\right) = 4.64 \pm 0.33 \text{ or } (4.31, 4.97).$$

10.65 (a) We first obtain:

$$\overline{x} = 8 \qquad \overline{y} = 5$$
$$\sum(x_i - \overline{x})^2 = 20 \qquad \sum(y_i - \overline{y})^2 = 14$$

Consequently, the pooled variance is given by

$$s^2_{\text{pooled}} = \frac{\sum(x_i - \overline{x})^2 + \sum(y_i - \overline{y})^2}{n_1 + n_2 - 2} = \frac{20 + 14}{5 + 4 - 2} = 4.857.$$

(b) The *t*-statistic is $T = \frac{(\overline{X} - \overline{Y}) - 2}{s_{\text{pooled}}\sqrt{\frac{1}{n_1} + \frac{1}{n_2}}}$ with d.f. $= n_1 + n_2 - 2$. In the present problem, we have $t = \frac{(8 - 5) - 2}{\sqrt{4.857}\sqrt{\frac{1}{5} + \frac{1}{4}}} = 0.6764$, with d.f. = 7.

10.67 A 95% confidence interval for $\mu_1 - \mu_2$ has the form

$$(\bar{x}_1 - \bar{x}_2) \pm t_{0.05}\sqrt{\frac{S_1^2}{n_1} + \frac{S_2^2}{n_2}} = (249 - 233) \pm 1.706\sqrt{\frac{19^2}{12} + \frac{45^2}{15}} = 16 \pm 21.919$$

or (-5.919, 37.919).

10.69 (a) Let μ_A = mean of City A and μ_B = mean of City B.

The question is whether or not $\mu_A \neq \mu_B$. So, we formulate the hypotheses:

$$H_0 : \mu_A - \mu_B = 0 \text{ versus } H_1 : \mu_A - \mu_B \neq 0$$

The sample sizes are $n_A = 75$ and $n_B = 100$ are large so we use the Z-test. The test statistic is $Z = \dfrac{\overline{X} - \overline{Y}}{\sqrt{\dfrac{S_A^2}{n_A} + \dfrac{S_B^2}{n_B}}}$. Use the sample data yields Z = -4.97. The associated p-value is $P(Z < -4.97) < 0.0001$, so we conclude that μ_A is significantly different than μ_B.

(b) A large sample 98% confidence interval for $\mu_A - \mu_B$ is given by

$$(\bar{x}_A - \bar{x}_B) \pm z_{0.02/2}\sqrt{\frac{s_A^2}{n_A} + \frac{s_B^2}{n_B}} = (37.8 - 43.2) \pm 2.56(1.086)$$

$$= -5.4 \pm 2.780$$

or equivalently (-5.399, -2.62).

(c) Large sample 98% confidence intervals for μ_A and μ_B separately are

$$\bar{x}_A \pm z_{0.02/2}\left(\frac{s_A}{\sqrt{n_A}}\right) = 37.8 \pm 2.56\left(\frac{6.8}{\sqrt{75}}\right) = 37.8 \pm 2.010,$$

or (35.79, 39.81)

$$\bar{x}_B \pm z_{0.02/2}\left(\frac{s_B}{\sqrt{n_A}}\right) = 43.2 \pm 2.56\left(\frac{7.5}{\sqrt{100}}\right) = 43.2 \pm 1.92,$$

or (41.28, 45.12)

10.71 (a) Since the p-value is 0.067, we do not reject H_0 at $\alpha = 0.05$.

(b) A large sample 95% confidence interval for $\mu_1 - \mu_2$ is given by

$$(\bar{y} - \bar{x}) \pm z_{0.05/2}\sqrt{\frac{s_1^2}{n_1} + \frac{s_2^2}{n_2}} = (2.945 - 2.643) \pm 1.96\sqrt{\frac{(0.554)^2}{15} + \frac{(0.390)^2}{20}}$$

$$= 0.302 \pm 1.96(0.1675) = 0.302 \pm 0.3283$$

or $(-0.026, 0.630)$.

(c) Denote by μ_1 and μ_2 the population mean CAS for female and male, respectively. In order to determine if μ_1 and μ_2 are significantly different, we test the hypotheses $H_0 : \mu_1 - \mu_2 = -0.1$ versus $H_1 : \mu_1 - \mu_2 \neq -0.1$ with $\alpha = 0.05$. We assume normal populations with equal variance, so the test statistic is

$$T = \frac{\overline{X} - \overline{Y}}{s_{\text{pooled}} \sqrt{\frac{1}{n_1} + \frac{1}{n_2}}} \text{ with d.f.} = n_1 + n_2 - 2$$

Since H_1 is two-sided and $\alpha = 0.05$, the rejection region is $R : |T| \geq t_{0.05/2} = 2.039$ (for d.f. = 15 + 20 – 2 = 33). From the sample data, we calculate the following:

$\overline{x} - \overline{y} = 2.643 - 2.945 = -0.302$

$$s_{\text{pooled}} = \sqrt{\frac{(n_1 - 1)s_1^2 + (n_2 - 1)s_2^2}{n_1 + n_2 - 2}} = \sqrt{\frac{14(0.554)^2 + 19(0.390)^2}{33}} = 0.467.$$

As such, the value of the observed t is

$$t = \frac{-0.302}{0.3416(0.467)} = -1.89,$$

which does not lie in R. So, H_0 is not rejected at $\alpha = 0.05$.

10.73 (a) The summary statistics are:

$$n_A = 8, \quad \overline{x} = 32, \quad s_A = 8.8318$$
$$n_B = 8, \quad \overline{y} = 27.5, \quad s_B = 8.3666$$

We test the hypotheses: $H_0 : \mu_A - \mu_B = 0$ versus $H_1 : \mu_A - \mu_B > 0$

We use the test statistic

$$T = \frac{\overline{X_A} - \overline{X_B}}{t_{0.05} \sqrt{\frac{S_A^2}{n_A} + \frac{S_B^2}{n_B}}}$$

with d.f. = the smaller of $n_A - 1$, $n_B - 1 = 7$. Since H_1 is right-sided, the rejection region is $R : T \geq t_{0.05}$. Since $n_1 = 8, n_2 = 8$, d.f.$= 7$, this region becomes $R : T \geq 1.805$. The value of the test statistic is

$$T = \frac{4.5}{1.805(4.301)} = 0.5797$$

Since this value is not in the rejection region R, we do not reject H_0; there is insufficient evidence to support the claim.

(b) The farms should be chosen from random locations within the state, and the plots within each farm should be chosen randomly.

(c) Yes, the data are just a scrambled version from 10.72. The results here are the same.

10.75 (a) This is a matched pair design. From the data on the paired differences $d=$ (lab A – lab B), we obtain the following summary statistics:

$$n=9,\ \bar{d}=9.667, \text{ and } s_D=14.5$$

We assume that the population distribution of these differences is normal, and denote the population mean by δ. We formulate the hypotheses

$$H_0:\delta=0 \text{ versus } H_1:\delta\neq 0.$$

The test statistic is $T=\dfrac{\bar{D}}{S_D/\sqrt{n}}$, d.f. = $n-1$. Since H_1 is two-sided and $\alpha=0.02$, the rejection region is $R:|T|\geq t_{0.01}=2.896$ (for d.f. = 8). Using the summary statistics, we see that the observed t is $t=\frac{9.667}{14.5/\sqrt{9}}=2.00$, which does not lie in R. Hence, H_0 is not rejected at $\alpha=0.02$, and so we conclude that the difference is not significant at this level.

(b) For $\alpha=0.10$ and d.f. = 8, we have $t_{0.05}=1.860$. So, the 90% confidence interval for δ is given by

$\bar{d}\pm t_{0.05}\frac{s_D}{\sqrt{n}}=9.667\pm 1.860\left(\frac{14.5}{\sqrt{9}}\right)=9.667\pm 8.999$ or $(0.6789, 18.654)$.

10.77 This is a matched pair design, where each location gives a pair. We calculate the paired differences $d=$ (with additive – without additive) as: $3,3,-1,-1,3,-1,4,2,4,4$. From this, we obtain the following summary statistics: $n=10,\ \bar{d}=2.0,$ and $s_D=2.160$

We assume that the population distribution of these differences is normal, and denote the population mean by δ. We formulate the hypotheses

$$H_0:\delta=0 \text{ versus } H_1:\delta>0.$$

The test statistic is $T=\dfrac{\bar{D}}{S_D/\sqrt{n}}$, d.f. = $n-1$. Since H_1 is right-sided and $\alpha=0.05$, the rejection region is $R:T\geq t_{0.05}=1.833$ (for d.f. = 9). Using the summary statistics, we see that the observed t is $t=\frac{2.0}{2.160/\sqrt{10}}=2.93$, which lies in R. Hence, H_0 is rejected at $\alpha=0.05$, and so we conclude that there is strong evidence that the additive is effective.

10.79 Denote by μ_1 and μ_2 the travel times for path A and path B, respectively. In order to determine if μ_1 and μ_2 are significantly different, we test the hypotheses $H_0:\mu_1-\mu_2=0$ versus $H_1:\mu_1-\mu_2\neq 0$ with $\alpha=0.05$. We assume normal populations with equal variance, so the test statistic is

$$T = \frac{\overline{X_1} - \overline{X_2}}{s_{\text{pooled}}\sqrt{\frac{1}{n_1} + \frac{1}{n_2}}} \text{ with d.f.} = n_1 + n_2 - 2$$

Since H_1 is two-sided and $\alpha = 0.05$, the rejection region is $R: |T| \geq t_{0.05/2} = 2.228$ (for d.f. $= 6 + 6 - 2 = 10$).

The summary statistics are:

$$n_1 = 6, \quad \overline{x_1} = 12.5, \quad s_1 = 2.429$$
$$n_2 = 6, \quad \overline{x_2} = 15.1, \quad s_2 = 2.317$$

From the sample data, we calculate the following:

$$\overline{x_1} - \overline{x_2} = 12.5 - 15.1 = -2.6$$

$$s_{\text{pooled}} = \sqrt{\frac{(n_1 - 1)s_1^2 + (n_2 - 1)s_2^2}{n_1 + n_2 - 2}} = \sqrt{\frac{5(2.429)^2 + 5(2.317)^2}{10}} = 2.374.$$

As such, the value of the observed t is

$$t = \frac{\overline{X_1} - \overline{X_2}}{s_{\text{pooled}}\sqrt{\frac{1}{n_1} + \frac{1}{n_2}}} = \frac{-2.6}{2.374(0.5774)} = -1.897,$$

which does not lie in R. So, H_0 is not rejected at $\alpha = 0.05$.

(b) Hold the condition fixed and collect all data on that particular day, to avoid the underlying condition from affecting the result.

10.81 Let p_1 and p_2 denote the probabilities of rain from seeded clouds and non-seeded clouds, respectively. We formulate the hypotheses:

$$H_0: p_1 = p_2 \text{ versus } H_1: p_1 < p_2.$$

The test statistic is $Z = \frac{\hat{p}_1 - \hat{p}_2}{\sqrt{\hat{p}\hat{q}}\sqrt{\frac{1}{n_1} + \frac{1}{n_2}}}$. We calculate the following:

$$\hat{p}_1 = \tfrac{7}{50} = 0.140, \quad \hat{p}_2 = \tfrac{43}{165} = 0.261$$

Pooled estimate $\hat{p} = \frac{n_1\hat{p}_1 + n_2\hat{p}_2}{n_1 + n_2} = \frac{50}{215} = 0.233,$

Observed z is $\frac{0.140 - 0.261}{\sqrt{(0.233)(0.767)}\sqrt{\frac{1}{50} + \frac{1}{165}}} = -1.77,$

The associated p-value is $P[Z \leq -1.77] = 0.0384$. So, H_0 would be rejected for α as small as 0.0384. As such, there is fairly strong evidence in support of the conjecture.

10.83 Let p_F and p_M denote the true proportion of females and males who have the antibody. We formulate the hypotheses:

$$H_0 : p_F = p_M \quad \text{versus} \quad H_1 : p_F > p_M .$$

The test statistic is $Z = \dfrac{\hat{p}_F - \hat{p}_M}{\sqrt{\hat{p}\hat{q}}\sqrt{\dfrac{1}{n_F}+\dfrac{1}{n_M}}}$, where $\hat{p}_F$ and $\hat{p}_M$ are identified with $\hat{p}_F$ and $\hat{p}_M$. Since H_1 is right-sided and $\alpha = 0.05$, the rejection region is $R : Z \geq z_{0.05/2} = 1.96$. Using the given sample statistics, we have

$$\text{Pooled estimate } \hat{p} = \frac{n_1\hat{p}_1 + n_2\hat{p}_2}{n_1 + n_2} = \frac{113(0.3009)+139(0.3885)}{113+139} = 0.3492,$$

$$\text{Observed } z \text{ is } \frac{0.3885-0.3009}{\sqrt{(0.3492)(0.6508)}\sqrt{\dfrac{1}{113}+\dfrac{1}{139}}} = 3.043,$$

lies in R. So, H_0 is rejected at $\alpha = 0.05$.

10.85 (a) Since the p-value is 0.127, which is greater than 0.05, there is not strong enough evidence to reject H_0.

(b) Enter the data into columns C3 and C4 of a Minitab worksheet. The output is as follows:

Two Sample T-Test and Confidence Interval

```
Two sample T for C3 vs C4

N        Mean       StDev    SE Mean
C3    8        73.9        10.1         3.6
C4   11        91.9        12.4         3.7

97% CI for mu C3 - mu C4: ( -30.3,   -5.8)
T-Test mu C3 = mu C4 (vs not =): T = -3.50
P = 0.0030 DF = 16
```

The 97% confidence interval is $(-30.3, -5.8)$.

10.87 (a) Denote by μ_1 and μ_2 the population mean for males at Lake Apopka and Lake Woodfruff, respectively. The summary statistics are:

Lake Apopka:	$n_1 = 5$,	$\bar{x} = 13.40$,	$s_1 = 8.82$
Lake Woodruff:	$n_2 = 9$,	$\bar{y} = 50.44$,	$s_2 = 27.13$

We should not pool the variances since the second standard deviation is more than 3 times the first.

(b) Using the conservative procedure, the 90% confidence interval for $\mu_1 - \mu_2$ has the form

$$\left(\overline{X}-\overline{Y}\right) \pm t_{\alpha/2}\sqrt{\frac{s_1^2}{n_1}+\frac{s_2^2}{n_2}}\text{ , d.f. = smaller of } n_1-1,\ n_2-1.$$

Using the sample statistics, we obtain:

$$\bar{x}-\bar{y}=13.40-50.44=-37.04$$

$$\sqrt{\frac{s_1^2}{n_1}+\frac{s_2^2}{n_2}}=\sqrt{\frac{(8.82)^2}{5}+\frac{(27.13)^2}{9}}=9.87$$

Since $t_{0.025}=2.132$ for d.f. = 4, the 90% confidence interval in this case is $-37.40 \pm (2.132)(9.87) = -37.40 \pm 21.04$ or $(-58.0, -16.0)$.

(c) The mean for males in the control group at Lake Woodruff is from 16 to 58 units larger than the mean for males at Lake Apopka.

10.89 Enter the data into columns C9 and C10 of a Minitab worksheet. The following is the output:

```
Two sample T for C9 vs C10

        N       Mean       StDev    SE Mean
C9     40      455.1        37.3        5.9
C10    40      429.1        41.0        6.5

95% CI for mu C9 - mu C10: ( 8.5,   43.4)
T-Test mu C9 = mu C10 (vs not =): T = 2.96  P = 0.0041  DF = 77
```

This was done without pooling the variance – note that since the ratio of the standard deviations is less than 1, one could have pooled the variance to get a slightly improved test statistic. Nonetheless, the *p*-value is 0.0041 with the more conservative estimate, so we reject H_0 and claim that the two means are different. Note also the 95% confidence interval is given.

10.91 Enter the data into two columns of a Minitab worksheet. The output is as follows:

```
Two sample T for Pre row vs Post row
             N       Mean      StDev     SE Mean
Pre row     81      725.9       89.3         9.9
Post row    81      673.3       83.0         9.2
95% CI for mu Pre row - mu Post row: ( 25.9,  79.4)
   T-Test mu Pre row = mu Post row (vs not =): T = 3.89
   P = 0.0001   DF = 159
```

There is a highly significant difference between mean pre row time and post row time.

Chapter 11

REGRESSION ANALYSIS I – SIMPLE LINEAR REGRESSION

11.1 The points on the line $y = 2 + 3x$ for $x = 1$ and $x = 4$ are (1,5) and (4,14) respectively. The intercept is 2 and the slope is 3.

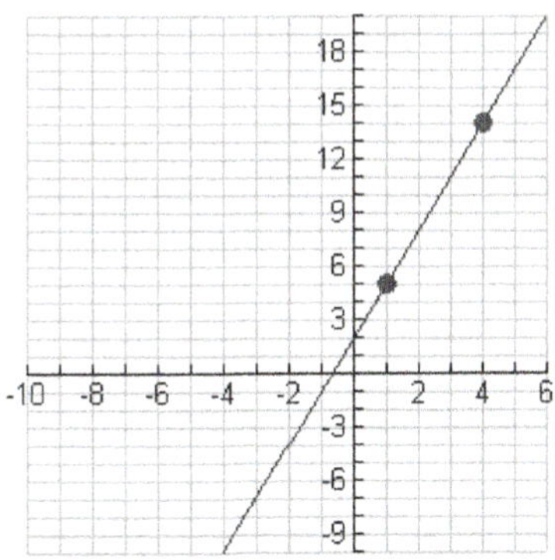

11.3 (a) Predictor variable x = duration of training
Response variable y = performance in skilled job

(b) Predictor variable x = average number of cigarettes smoked daily
Response variable y = CO level in blood

(c) Predictor variable x = Humidity level in environment
Response variable y = Growth rate of fungus

(d) Predictor variable x = Expenditures in promoting product
Response variable y = Amount of product sales

11.5 The model is $Y = \beta_0 + \beta_1 x + e$, where $E(e) = 0$ and $sd(e) = \sigma$, so $\beta_0 = 6,\ \beta_1 = -3,$ and $\sigma = 3$.

11.7 $Y = \beta_0 + \beta_1 x + e = 3 - 2x + e$, where $E(e) = 0$ and $sd(e) = \sigma$.

(a) At $x = 1$, $E(Y) = 3 - 2(1) = 1$ and $sd(Y) = sd(e) = 3$.

(b) At $x = 2$, $E(Y) = 3 - 2(2) = -1$ and $sd(Y) = sd(e) = 3$.

11.9 The straight line for the means of the model $Y = 7 + 2x + e$ is $y = 7 + 2x$. The graph of the line is shown below.

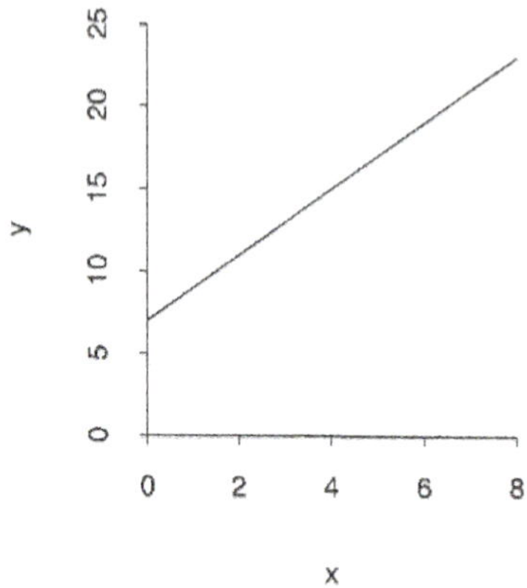

11.11 (a) At $x = 4$, $E(Y) = \beta_0 + \beta_1(4) = 4 + 3(4) = 16$.

At $x = 5$, $E(Y) = \beta_0 + \beta_1(5) = 4 + 3(5) = 19$.

(b) No, only the mean is larger. By chance the error e at $x = 5$, which has standard deviation 4, could be quite negative and/or the error at $x = 4$ very large.

11.13 (a) The scatter diagram is shown below.

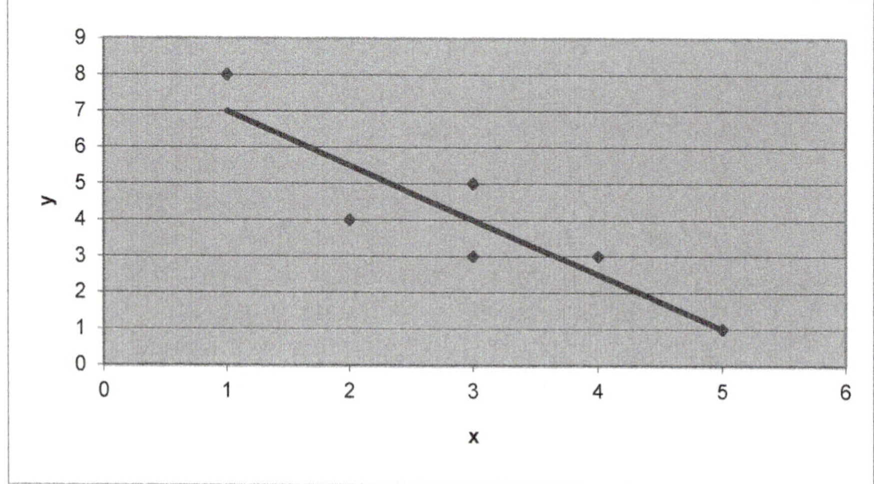

(b) The computations needed to calculate $\bar{x}$, $\bar{y}$, S_{xx}, S_{xy}, and S_{yy} are provided in the following table:

	x	y	$x-\bar{x}$	$y-\bar{y}$	$(x-\bar{x})(y-\bar{y})$	$(x-\bar{x})^2$	$(y-\bar{y})^2$
	1	8	−2	4	-8	4	16
	2	4	−1	0	0	1	0
	3	5	0	1	0	0	1
	3	3	0	-1	0	0	1
	4	3	1	-1	-1	1	1
	5	1	2	-3	-6	4	9
Total	18	24	0	0	-15	10	28

So, we have

$$\bar{x} = \frac{\sum x}{n} = \frac{18}{6} = 3 \qquad S_{xx} = \sum(x-\bar{x})^2 = 10$$

$$\bar{y} = \frac{\sum y}{n} = \frac{24}{6} = 4 \qquad S_{xy} = \sum(x-\bar{x})(y-\bar{y}) = -15$$

$$S_{yy} = \sum(y-\bar{y})^2 = 28$$

(c) $\hat{\beta}_1 = \frac{S_{xy}}{S_{xx}} = -\frac{15}{10} = -1.5, \quad \hat{\beta}_0 = \bar{y} - \hat{\beta}_1\bar{x} = 4 - (-1.5)(3) = 8.5$

(d) The fitted line is $\hat{y} = 8.5 - 1.5x$, which is graphed in part (a).

11.15 (a) The residuals and their sum are calculated in the following table:

	x	y	$\hat{y} = 8.5 - 1.5x$	$\hat{e} = y - \hat{y}$	$(y-\hat{y})^2$
	1	8	7	1	1
	2	4	5.5	-1.5	2.25
	3	5	4	1	1
	3	3	4	-1	1
	4	3	2.5	0.5	0.25
	5	1	1	0	0
Total					5.50

So, $\sum(y-\hat{y}) = 0.0$.

(b) SSE = Sum of squares residuals = 5.50

$\text{SSE} = S_{yy} - \frac{S_{xy}^2}{S_{xx}} = 28 - \frac{(-15)^2}{10} = 5.50$ (Check this using the calculations in Exercise 11.13.)

(c) $S^2 = \frac{\text{SSE}}{n-2} = \frac{5.50}{6-2} = 1.375$

11.17 (a) The computations needed to calculate $\bar{x}$, $\bar{y}$, S_{xx}, S_{xy}, and S_{yy} are provided in the following table:

	x	y	$x-\bar{x}$	$y-\bar{y}$	$(x-\bar{x})(y-\bar{y})$	$(x-\bar{x})^2$	$(y-\bar{y})^2$
	1	4	–3	–2	6	9	4
	2	3	–2	–3	6	4	9
	4	6	0	0	0	0	0
	6	8	2	2	4	4	4
	7	9	3	3	9	9	9
Total	20	30	0	0	25	26	26

So, we have

$$\bar{x} = \frac{\sum x}{n} = \frac{20}{5} = 4 \qquad S_{xx} = \sum(x-\bar{x})^2 = 26$$

$$\bar{y} = \frac{\sum y}{n} = \frac{30}{5} = 6 \qquad S_{xy} = \sum(x-\bar{x})(y-\bar{y}) = 25$$

$$S_{yy} = \sum(y-\bar{y})^2 = 26$$

(b) $\hat{\beta}_1 = \dfrac{S_{xy}}{S_{xx}} = \dfrac{25}{26} = 0.96, \quad \hat{\beta}_0 = \bar{y} - \hat{\beta}_1\bar{x} = 6-(0.96)(4) = 2.16$

(c) The fitted line is $\hat{y} = 2.16 - 0.96x$.

(d) $\hat{y} = 2.16 - 6.5(0.96) = -4.08$

11.19 (a)

$$\hat{\beta}_1 = \frac{S_{xy}}{S_{xx}} = \frac{19.305}{1.607} = 12.013,$$

$$\hat{\beta}_0 = \bar{y} - \hat{\beta}_1\bar{x} = 33.416 - (12.013)(2.526) = 3.071$$

The fitted line is $\hat{y} = 3.071 + 12.013x$.

(b) $\text{SSE} = S_{yy} - \dfrac{S_{xy}^2}{S_{xx}} = 448.839 - \dfrac{(19.305)^2}{1.607} = 216.927$

(c) $\sigma^2 \approx S^2 = \dfrac{\text{SSE}}{n-2} = \dfrac{216.927}{12} = 18.077$

11.21 We first calculate the means and sums of squares and products:

$$\bar{y} = \frac{\sum y}{n} = \frac{889}{7} = 127, \qquad \bar{x} = \frac{\sum x}{n} = \frac{520}{7} = 74.286$$

$$\sum_{i=1}^{7} y_i^2 = 113,237, \quad \sum_{i=1}^{7} x_i^2 = 39,328, \quad \sum_{i=1}^{7} x_i y_i = 66,392$$

Thus,

$$S_{xx} = \sum_{i=1}^{7} x_i^2 - \left(\sum_{i=1}^{7} x_i\right)^2 / 7 = 39,328 - (520)^2 / 7 = 699.43$$

$$S_{xy} = \sum_{i=1}^{7} x_i y_i - \left(\sum_{i=1}^{7} x_i\right)\left(\sum_{i=1}^{7} y_i\right) / 7 = 66,392 - (520)(889)/7 = 352$$

$$S_{yy} = \sum_{i=1}^{7} y_i^2 - \left(\sum_{i=1}^{7} y_i\right)^2 / 7 = 113,237 - (889)^2 / 7 = 334$$

(a) $\hat{\beta}_1 = \dfrac{S_{xy}}{S_{xx}} = \dfrac{352}{699.43} = 0.5033,$

$\hat{\beta}_0 = \bar{y} - \hat{\beta}_1\bar{x} = 127 - (0.5033)(74.286) = 89.61$

The fitted line is $\hat{y} = 89.61 + 0.5033x$.

(b) $\text{SSE} = S_{yy} - \dfrac{S_{xy}^2}{S_{xx}} = 334 - \dfrac{(352)^2}{699.43} = 156.85$

(c) $S^2 = \dfrac{\text{SSE}}{n-2} = \dfrac{156.85}{5} = 31.37$

(d) No. The line with x as a predictor is not the inverse of that with y as predictor because the least squares problem is not a linear one.

11.23 At $\bar{x}$, $\hat{y} = \hat{\beta}_0 + \hat{\beta}_1\bar{x} = \underbrace{\bar{y} - \hat{\beta}_1\bar{x}}_{=\hat{\beta}_0} + \hat{\beta}_1\bar{x} = \bar{y}$

11.25 (a) The computations needed to calculate $\bar{x}$, $\bar{y}$, S_{xx}, S_{xy}, and S_{yy} are provided in the following table:

	x	y	$x-\bar{x}$	$y-\bar{y}$	$(x-\bar{x})(y-\bar{y})$	$(x-\bar{x})^2$	$(y-\bar{y})^2$
	1	5	-2	-6	12	4	36
	2	11	-1	0	0	1	0
	3	9	0	-2	0	0	4
	4	14	1	3	3	1	9
	5	16	2	5	10	4	25
Total	15	55	0	0	25	10	74

So, we have

$$\bar{x} = \tfrac{\sum x}{n} = \tfrac{15}{5} = 3 \qquad S_{xx} = \sum (x-\bar{x})^2 = 10$$

$$\bar{y} = \tfrac{\sum y}{n} = \tfrac{55}{5} = 11 \qquad S_{xy} = \sum (x-\bar{x})(y-\bar{y}) = 25$$

$$S_{yy} = \sum (y-\bar{y})^2 = 74$$

$$\hat{\beta}_1 = \frac{S_{xy}}{S_{xx}} = \frac{25}{10} = 2.5, \qquad \hat{\beta}_0 = \bar{y} - \hat{\beta}_1\bar{x} = 11 - (2.5)(3) = 3.5$$

$$\text{SSE} = S_{yy} - \frac{S_{xy}^2}{S_{xx}} = 74 - \frac{(25)^2}{10} = 11.5$$

$$S^2 = \frac{\text{SSE}}{n-2} = \frac{11.5}{3} = 3.833$$

(b) We test the hypotheses: $H_0 : \beta_1 = 0$ versus $H_1 : \beta_1 \neq 0$
Since H_1 is two –sided and $\alpha = 0.05$, the rejection region is $R: |T| > t_{0.025} = 3.182$ (for d.f. = 3). The value of the observed t is

$$t = \frac{\hat{\beta}_1 - 0}{s / \sqrt{S_{xx}}} = \frac{2.5}{\sqrt{3.833/10}} = 4.038,$$

which lies in R. Hence, H_0 is rejected at $\alpha = 0.05$.

(c) The expected value is estimated by $3.5 + 2.5(3) = 11$. Since

$$s\sqrt{\frac{1}{n} + \frac{(3-\bar{x})^2}{S_{xx}}} = \sqrt{3.833}\sqrt{\frac{1}{5} + \frac{(3-3)^2}{10}} = 0.8756$$

and the upper 0.05 point of the t with d.f. = 3 is 2.353, the 90% confidence interval for the expected y value is given by

$$11 \pm 2.353(0.8756) \quad \text{or} \quad (8.9397, 13.0603).$$

11.27 Since $t_{0.025} = 3.182$ for d.f. = 3, a 95% confidence interval for β_1 is given by

$$\hat{\beta}_1 \pm 3.182 \frac{s}{\sqrt{S_{xx}}} = 2.5 \pm 3.182\sqrt{\frac{3.833}{10}} = 2.5 \pm 1.970$$

or $(0.53, 4.47)$.

11.29 (a) & (b) Using Minitab, we have:

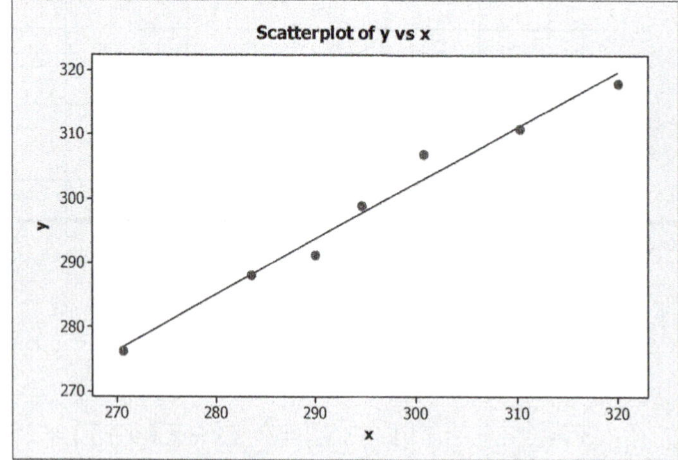

Note the equation of the least squares regression line is $y = 0.8694x + 41.58$. This could have been computed by hand using the same calculations used in exercises from the previous section.

(c) First, note that $t_{0.025} = 2.571$ for d.f. = 5. In order to determine the confidence interval for $\hat{\beta}_1$, we need the following computations:

	x	y	$x-\bar{x}$	$y-\bar{y}$	$(x-\bar{x})(y-\bar{y})$	$(x-\bar{x})^2$	$(y-\bar{y})^2$
	283.5	288	−12.2	−10.6	129.32	148.84	112.36
	290	291.2	−5.7	−7.4	42.18	32.49	54.76
	270.5	276.2	−25.2	−22.4	564.48	635.04	501.76
	300.8	307	5.1	8.4	42.84	26.01	70.56
	310.2	311	14.5	12.4	179.80	210.25	153.76
	294.6	299	−1.1	0.4	−0.44	1.21	0.16
	320	318	24.3	19.4	471.40	590.49	376.36
Total	2069.6	2090.4			1429.6	1644.3	1269.7

So, we have

$$\bar{x} = \tfrac{\sum x}{7} = 295.7 \qquad S_{xx} = \sum(x-\bar{x})^2 = 1644.3$$

$$\bar{y} = \tfrac{\sum y}{7} = 298.6 \qquad S_{xy} = \sum(x-\bar{x})(y-\bar{y}) = 1429.6$$

$$S_{yy} = \sum(y-\bar{y})^2 = 1269.7$$

Thus, we have

$$\text{SSE} = S_{yy} - \frac{S_{xy}^2}{S_{xx}} = 26.766 \qquad S^2 = \frac{\text{SSE}}{n-2} = \frac{26.766}{5} = 5.3532.$$

Therefore, $\dfrac{s}{\sqrt{S_{xx}}} = \dfrac{2.314}{\sqrt{1644.3}} = 0.0571$. Also, $\hat{\beta}_1 = \dfrac{S_{xy}}{S_{xx}} = 0.8694$.

So, a 95% confidence interval for β_1 is given by

$$\hat{\beta}_1 \pm 2.571\frac{s}{\sqrt{S_{xx}}} = 0.8694 \pm 2.571(0.0571) = 0.8694 \pm 0.1468$$

or $(0.7226,\ 1.016)$.

11.31 (a) The expected value of HDI corresponding to $x^* = 22$ internet users per 100 is estimated as

$$\widehat{\beta_0} + \widehat{\beta_1}x^* = 0.4242 + 0.0062(22) = 0.5606$$

Its estimated SE = $0.0263\sqrt{\dfrac{1}{15} + \dfrac{(22-30.133)^2}{8062.39}} = 0.2736$

Since $t_{0.025} = 2.160$ for d.f. = 13, the 95% confidence interval is

$$0.5606 \pm 2.160(0.2736) = 0.5606 \pm 0.5910$$

or (-0.0304, 1.152).

The width of the confidence interval in Example 9 is 0.103, whereas the interval just computed has width 1.1824.

(b) Now, we compute the 95% confidence interval for a single country corresponding to $x^* = 22$. As in (a), $\widehat{\beta_0} + \widehat{\beta_1} x^* = 0.5606$, but the estimated SE is now

$$0.0263\sqrt{1 + \frac{1}{15} + \frac{(22 - 30.133)^2}{8062.39}} = 0.0384$$

Since $t_{0.025} = 2.160$ for d.f. = 13, the 95% confidence interval is

$$0.5606 \pm 2.160(0.0384) = 0.5606 \pm 0.0829$$

or (0.4777, 0.6435).

(c) No, it cannot establish causality.

11.33 (a) The model is $Y = \beta_0 + \beta_1 x + e$ and the fit suggested by the data is

$$\hat{y} = 994 + 0.10373x \text{ with } \widehat{sd(e)} = \hat{\sigma} = 299.4.$$

Note that the r^2 is only 0.302. This means that only 30.2% of the variability in the data is explained by the model (refer to Exercise 11.44).

(b) The t-ratio on the computer output is the t-statistic for testing that the coefficient is zero. Since the t-ratio for the x term is 3.48 with p-value 0.002, we reject H_0: $\beta_0 = 0$ at $\alpha = 0.05$.

11.35 (a) The model is $Y = \beta_0 + \beta_1 x + e$ and the fit suggested by the data is

$$\hat{y} = 0.3381 + 0.83099x \text{ with } \widehat{sd(e)} = \hat{\sigma} = 0.1208.$$

(b) Since the t-ratio for the x term is 9.55 with p-value less than 0.0001, we reject H_0: $\beta_1 = 0$ at $\alpha = 0.05$. As such, the x term is needed in the model.

11.37 (a) Using Minitab, we find that the fitted line plot is as follows:

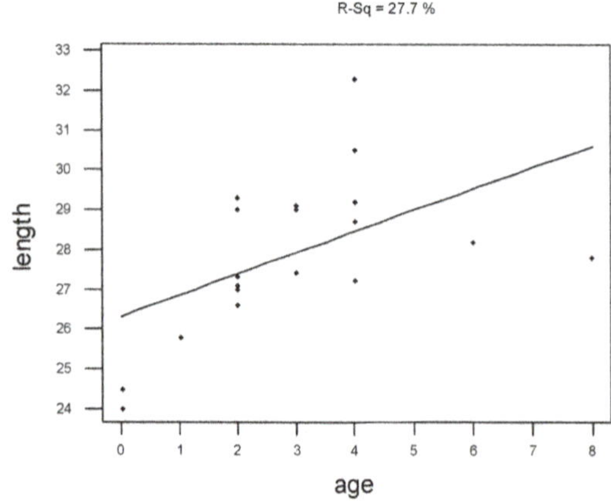

(b) Enter the data into a Minitab worksheet. The output is as follows:

Regression Analysis

```
The regression equation is
length = 26.3 + 0.538 age

Predictor        Coef       StDev          T        P
Constant      26.3101      0.7356      35.77    0.000
age            0.5377      0.2105       2.55    0.021

S = 1.722        R-Sq = 27.7%      R-Sq(adj) = 23.5%

Analysis of Variance

Source          DF          SS           MS          F         P
Regression       1      19.353       19.353       6.52     0.021
Residual Error      17       50.437        2.967
Total               18       69.789
```

Look in the age row – the *p*-value of 0.021 is the result of the hypothesis test of the slope at the 95% level. Here, we reject H_0 in favor of claiming there is linear relationship between the two variables.

(c) & (d) One can proceed by hand as in other exercises/examples. We, however choose to use Minitab to obtain the following two 90% confidence intervals corresponding to age $x = 4$:

```
Predicted Values
Fit    StDev Fit             90.0% CI                  90.0% PI
28.461       0.453     (  27.673,   29.249)   (  25.362, 31.559)
```

11.39 (a) $r^2 = \dfrac{S_{xy}^2}{S_{xx}S_{yy}} = 0.517$. So, about 51.7%. (b) Same as (a).

(c) The best fit line will be different, but the value of r^2 will be the same.

11.41 (a) and (b): The r^2 value is the same as in Exercise 11.40.

11.43 By Exercise 11.28, we have $S_{xx} = 10,\ S_{yy} = 0.82,\ S_{xy} = 2.8$. So,

(a) Proportion of variance explained $r^2 = \dfrac{S_{xy}^2}{S_{xx}S_{yy}} = 0.956$

(b) $r = \dfrac{S_{xy}}{\sqrt{S_{xx}S_{yy}}} = 0.978$

11.45 Proportion explained $= r^2 = 0.799$

11.47 (a) Recall $\hat{\beta}_1 = \frac{S_{xy}}{S_{xx}}$, so multiplying r by $\frac{S_{xx}}{S_{xx}}$

$$r = \frac{S_{xy}}{\sqrt{S_{xx}S_{yy}}} = \frac{S_{xx}}{S_{xx}}\frac{S_{xy}}{\sqrt{S_{xx}}}\frac{1}{\sqrt{S_{yy}}} = \hat{\beta}_1 \frac{S_{xx}}{\sqrt{S_{xx}}}\frac{1}{\sqrt{S_{yy}}} = \hat{\beta}_1 \frac{\sqrt{S_{xx}}}{\sqrt{S_{yy}}}$$

(b) $$\text{SSE} = S_{yy} - \frac{S_{xy}^2}{S_{xx}} = S_{yy} - \frac{S_{xy}^2}{S_{xx}}\frac{S_{yy}}{S_{yy}} = S_{yy}\left(1 - \frac{S_{xy}^2}{S_{xx}S_{yy}}\right) = S_{yy}(1-r^2)$$

11.49 The product x = (leaf length) × (leaf width) is the area of a rectangle that contains the leaf. It should be larger than the leaf, so the slope should be less than one.

11.51 (a) Note that

$$\hat{\beta}_1 = \frac{S_{xy}}{S_{xx}} = 3.144$$

$$\hat{\beta}_0 = \bar{y} - \beta_1\bar{x} = 12 - 3.144(2.556) = 3.964$$

So, $\hat{y} = \hat{\beta}_0 + \hat{\beta}_1 x$.

The residuals and their sum are calculated in the following table:

	x	y	$\hat{y}$	$\hat{e} = y - \hat{y}$	$(y-\hat{y})^2$
	1	8	7.108	1.108	1.228
	1	6	7.108	-1.108	1.228
	1	7	7.108	-0.108	0.012
	2	10	10.252	-0.252	0.064
	3	15	13.396	1.604	2.573
	3	12	13.396	-1.396	1.949
	3	13	13.396	-0.396	0.157
	4	19	16.54	2.46	6.052
	5	18	19.684	-1.684	2.836
Total					16.099

So, $\text{SSE} = 16.099$.

(b) SSE = Sum of squares residuals = 16.099

$$\text{SSE} = S_{yy} - \frac{S_{xy}^2}{S_{xx}} = 176 - \frac{51^2}{16.222} = 16.099$$

(c) $$S^2 = \frac{\text{SSE}}{n-2} = \frac{16.099}{9-2} = 2.300$$

11.53 (a) $\hat{\beta}_1 = \frac{S_{xy}}{S_{xx}} = -\frac{12.4}{5.6} = -2.214, \quad \hat{\beta}_0 = \bar{y} - \hat{\beta}_1\bar{x} = 54.8 + (2.214)(8.3) = 73.18$

The fitted line is then given by $\hat{y} = 73.18 - 2.214x$.

Also, $\text{SSE} = S_{yy} - \frac{S_{xy}^2}{S_{xx}} = 38.7 - \frac{(12.4)^2}{5.6} = 11.24$

$$S^2 = \frac{\text{SSE}}{n-2} = \frac{11.24}{13} = 0.8648$$

(b) We test the hypotheses: $H_0 : \beta_1 = -2$ versus $H_1 : \beta_1 < -2$
Since H_1 is left-sided and $\alpha = 0.05$, the rejection region is $R: T < -t_{0.05} = -1.771$ (for d.f. = 13). The value of the observed t is

$$t = \frac{\hat{\beta}_1 - (-2)}{s/\sqrt{S_{xx}}} = \frac{-2.214 + 2}{\sqrt{0.8648/5.6}} = -0.5446,$$

which does not lie in R. Hence, H_0 is not rejected at $\alpha = 0.05$.

(c) A 95% confidence interval (for $x^* = 10$) of the expected response 51.04 is given by

$$51.04 \pm 2.160(0.93)\sqrt{\frac{1}{15} + \frac{(10-8.3)^2}{5.6}} \quad \text{or} \quad (49.51, 52.57).$$

11.55 (a) The scatter diagram is given by

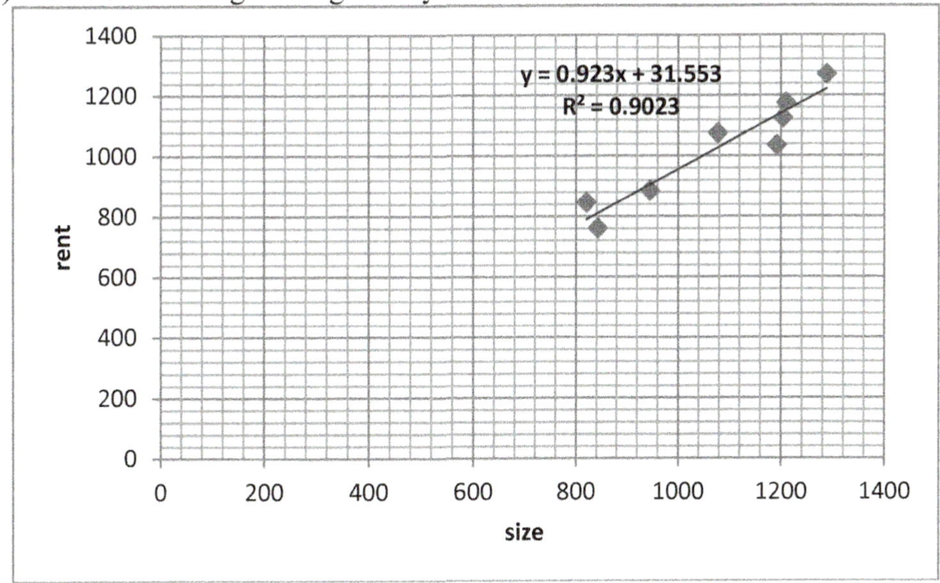

We first calculate the means and sums of squares and products:

$$\bar{x} = \frac{\sum x}{8} = 1072.25, \qquad \bar{y} = \frac{\sum y}{8} = 1021.25$$

$$\sum_{i=1}^{8} x_i^2 = 9427842, \quad \sum_{i=1}^{8} y_i^2 = 8560850, \quad \sum_{i=1}^{8} x_i y_i = 8972650$$

Thus,

$$S_{xx} = \sum_{i=1}^{8} x_i^2 - \left(\sum_{i=1}^{8} x_i\right)^2 / 8 = 230081.5$$

$$S_{xy} = \sum_{i=1}^{8} x_i y_i - \left(\sum_{i=1}^{8} x_i\right)\left(\sum_{i=1}^{8} y_i\right) / 8 = 212367.5$$

$$S_{yy} = \sum_{i=1}^{8} y_i^2 - \left(\sum_{i=1}^{8} y_i\right)^2 / 8 = 217237.5$$

Consequently, we have

$$\hat{\beta}_1 = \frac{S_{xy}}{S_{xx}} = 0.92301, \quad \hat{\beta}_0 = \bar{y} - \hat{\beta}_1 \bar{x} = 31.5525$$

The fitted line is $\hat{y} = 31.5525 + 0.92301x$.
Furthermore,

$$\text{SSE} = S_{yy} - \frac{S_{xy}^2}{S_{xx}} = 21220.197$$

$$S^2 = \frac{\text{SSE}}{n-2} = \frac{21220.197}{6} = 3536.70, \text{ so that } S = 59.47.$$

(b) We test the hypotheses: $H_0 : \beta_1 = 0$ versus $H_1 : \beta_1 > 0$
Since H_1 is right-sided and $\alpha = 0.05$, the rejection region is $R: T > t_{0.05} = 1.943$ (for d.f. = 6). The value of the observed t is

$$t = \frac{\hat{\beta}_1 - 0}{s / \sqrt{S_{xx}}} = \frac{0.92301}{59.47 / \sqrt{230081.5}} = 7.445,$$

which lies in R. Hence, H_0 is rejected at $\alpha = 0.05$. This implies that the mean rent y increases with size x.

(c) The expected increase is β_1. A 95% confidence interval is calculated as

$$0.92301 \pm 2.447\left(\frac{59.47}{\sqrt{230081.5}}\right) = 0.92301 \pm 0.30338 \quad \text{or} \quad (0.6196,\ 1.2264).$$

(d) For a specific apartment of size $x = 1025$, a 95% prediction interval is given by

$$[31.5525 + 0.92301(1025)] \pm 2.447(59.47)\sqrt{1 + \frac{1}{8} + \frac{(1025 - 1072.25)^2}{230081.5}}$$

$$= 977.638 \pm 155.015$$

or (822.623, 1132.653).

11.57 (a) & (b) The scatter diagram and fitted line are shown below:

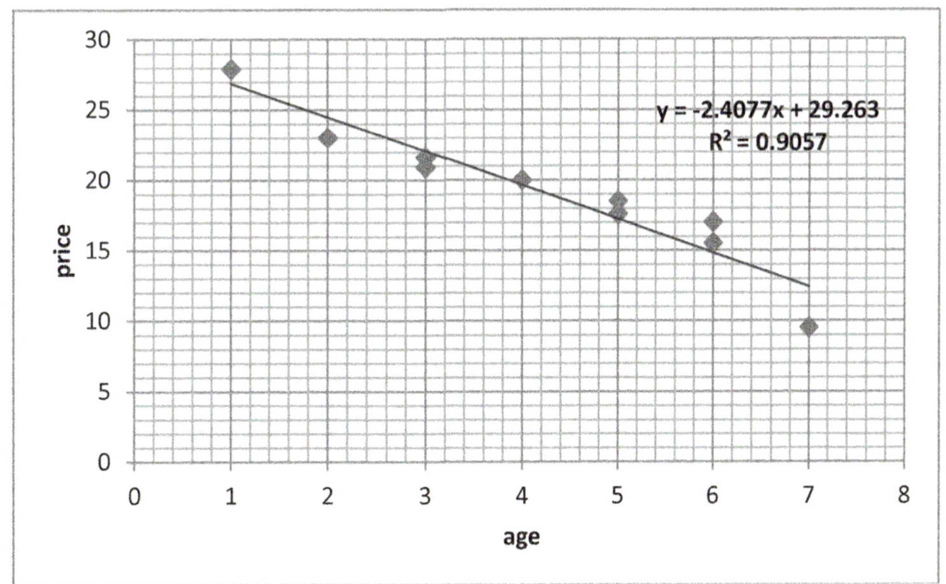

$\bar{x} = 4.2 \qquad S_{xx} = \sum (x-\bar{x})^2 = 33.6$

$\bar{y} = 19.15 \qquad S_{xy} = \sum (x-\bar{x})(y-\bar{y}) = -80.9$

$S_{yy} = \sum (y-\bar{y})^2 = 215.065$

As such, we have

$$\hat{\beta}_1 = \frac{S_{xy}}{S_{xx}} = -2.408, \quad \hat{\beta}_0 = \bar{y} - \beta_1\bar{x} = 29.26$$

So, the fitted line is $\hat{y} = 29.26 - 2.408x$.

(c) Note that $\text{SSE} = S_{yy} - \frac{S_{xy}^2}{S_{xx}} = 20.28$, so that $s = \sqrt{\frac{SSE}{8}} = 1.592$.

Since $t_{0.025} = 2.306$ for d.f. = 8, a 95% confidence interval for β_1 is

$$\hat{\beta}_1 \pm 2.306\frac{s}{\sqrt{S_{xx}}} = -2.408 \pm 2.306\left(\frac{1.592}{\sqrt{33.6}}\right) = -2.408 \pm 0.633$$

or (-3.041, -1.775).

11.59 $r = \frac{S_{xy}}{\sqrt{S_{xx}S_{yy}}} = \frac{-80.9}{\sqrt{(33.6)(215.065)}} = -0.95169$. Since $r^2 = 0.9057$ is the proportion of variance explained by the linear regression of y on x, the fit appears to be inadequate.

11.61

(a) $r = 0.649$ and $r^2 = 0.421$

(b) $r = 0.279$ and $r^2 = 0.078$

(c) $r = 0.733$ and $r^2 = 0.537$

(d) The pattern is quite different for male and female wolves. The scatter diagram is as follows:

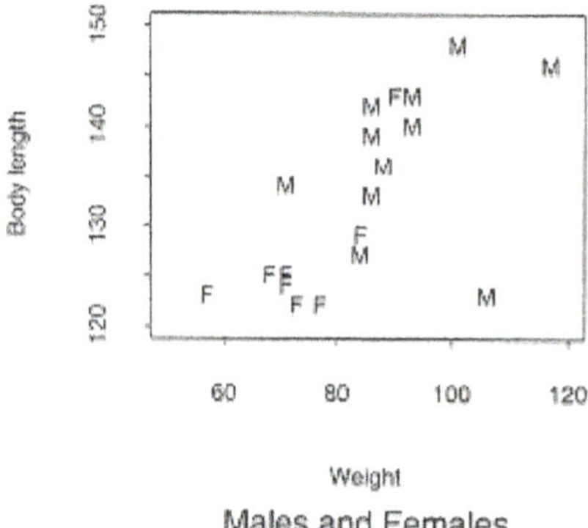

11.63

The Minitab output is on the next page.

(a) From the output, the fitted line is $\hat{y} = -87.17 + 1.2765x$.

(b) For d.f. = 16, $t_{0.05} = 1.746$. Therefore, the null hypothesis $H_0 : \beta_1 = 0$ will be rejected at $\alpha = 0.05$ if the observed t value is in the rejection region $R : T \geq 1.746$. According to the output, the observed t is given by $t = \frac{1.2765}{0.2156} = 5.92$, which lies in R. Hence, H_0 is rejected at $\alpha = 0.05$. Furthermore, since the p-value is very small, the data strongly support that $\beta_1 > 0$ which, in turn, indicates that the expected value of weight increases with body length.

(c) The intercept term is now needed since the p-value is. 0.008 The estimated slope has increased.

```
The regression equation is
weight0 = - 87.2 + 1.28 bodylen0

Predictor        Coef      StDev          T       P
Constant       -87.17      28.82      -3.02   0.008
bodylen0       1.2765     0.2156       5.92   0.000

S = 7.980         R-Sq = 68.7%      R-Sq(adj) = 66.7%

Analysis of Variance

Source           DF       SS       MS       F       P
Regression        1   2233.0   2233.0   35.07   0.000
Residual Error   16   1018.8     63.7
Total            17   3251.8

Unusual Observations
Obs bodylen0 weight0   Fit StDev Fit Residual St Resid
  8      146  117.00 99.21      3.31    17.79    2.45R
```

11.65 (a) The data for salmon growth are entered into columns C1 and C4 of a Minitab worksheet. And, the data for all salmon growth are stacked into C5 and C6. From the output (seen below), the freshwater growth of salmon is not an effective predictor for its marine growth.

```
STACK (C1 C2) (C3 C4) INTO C5 C6
NAME C5 'FRESHGRW' C6 'MARINGRW'
REGRESS C6 ON 1 PREDICTOR IN C5

THE REGRESSION EQUATION IS
MARINGRW = 470 - 0.258 FRESHGRW
PREDICTOR        COEF        STDEV     T-RATIO      P
CONSTANT       469.62        18.32       25.64   0.000
FRESHGRW      -0.2579       0.1664       -1.55   0.125

S = 40.72        R-SQ = 3.0%       R-SQ(ADJ) = 1.7%

ANALYSIS OF VARIANCE

SOURCE        DF        SS        MS       F       P
REGRESSION     1      3981      3981    2.40   0.125
ERROR         78    129323      1658
TOTAL         79    133305

CORR C1 C2

CORRELATION OF C1 AND C2 = -0.191
```

(b) The data for male salmon growth are entered into columns C1 and C2. From the output (seen below), the freshwater growth of a male salmon is not an effective predictor of its marine growth.

```
NAME C1 'FRESHGRW' C2 'MARINGRW'
REGRESS C2 ON 1 PREDICTOR IN C1

THE REGRESSION EQUATION IS
MARINGRW = 478 - 0.236 FRESHGRW

PREDICTOR        COEF        STDEV     T-RATIO      P
CONSTANT       478.35        20.30       23.57   0.000
FRESHGRW      -0.2364       0.1976       -1.20   0.239

S = 37.05        R-SQ = 3.6%       R-SQ(ADJ) = 1.1%

ANALYSIS OF VARIANCE

SOURCE        DF        SS        MS       F       P
REGRESSION     1      1966      1966    1.43   0.239
ERROR         38     52168      1373
TOTAL         39     54134

CORR C3 C4

CORRELATION OF C3 AND C4 = 0.040
```

(c) The data for female salmon growth are entered into columns C3 and C4. From the output (seen on the next page), the freshwater growth of a female salmon is not an effective predictor for its marine growth.

```
NAME C3 'FRESHGRW' C4 'MARINGRW'
REGRESS C4 ON 1 PREDICTOR IN C3

THE REGRESSION EQUATION IS
MARINGRW = 421 + 0.074 FRESHGRW

PREDICTOR      COEF      STDEV    T-RATIO      P
CONSTANT     420.63      35.00      12.02   0.000
FRESHGRW     0.0742     0.2994       0.25   0.806

S = 41.55      R-SQ = 0.2%     R-SQ(ADJ) = 0.0%

ANALYSIS OF VARIANCE

SOURCE        DF        SS          MS        F       P
REGRESSION     1       106         106     0.06   0.806
ERROR         38     65597        1726
TOTAL         39     65703

     CORR C5 C6

CORRELATION OF C5 AND C6 = -0.173
```

11.67 (a) Use Minitab to compute the proportion of variation in speed due to regression is `R-Sq = 88.6%`

(b) The Minitab output is as follows:

Regression Analysis: speed versus height

```
The regression equation is speed = 33.5 + 0.193 height

Predictor      Coef  SE Coef     T      P
Constant     33.453    7.095  4.71  0.001
height      0.19291  0.02184  8.83  0.000

S = 5.85513   R-Sq = 88.6%   R-Sq(adj) = 87.5%

Analysis of Variance

Source          DF      SS      MS      F      P
Regression       1  2675.8  2675.8  78.05  0.000
Residual Error  10   342.8    34.3
Total           11  3018.7

Unusual Observations

Obs  height   speed     Fit  SE Fit  Residual  St Resid
  3     415  100.00  113.51    2.75    -13.51    -2.61R
 R denotes an observation with a large standardized residual.
```

Since the regression equation is y = 33.5 + 0.193x, if $x = 425$, then the top predicted speed is $y = 33.5 + 0.193(425) = 115.5$.

(c) Using the same line, if x =,490 then the top predicted speed is 128.07 . Since the value of x is getting quite far away from the x-values of known data points, the predictive power of the regression line is severely diminished.

Chapter 12

REGRESSION ANALYSIS II – MULTIPLE LINEAR REGRESSION AND OTHER TOPICS

12.1 (a) The scatter diagram and fitted line are illustrated below:

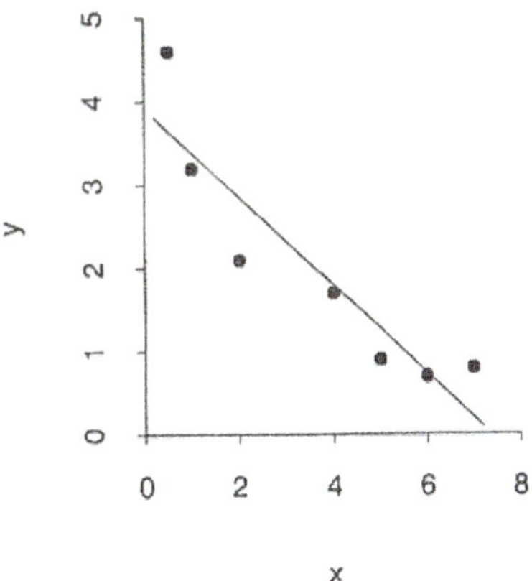

(b) Observe that

$\bar{x} = \frac{\sum x}{n} = \frac{25.5}{7} = 3.6429$ $\qquad S_{xx} = \sum (x-\bar{x})^2 = 38.3571$

$\bar{y} = \frac{\sum y}{n} = \frac{14.0}{7} = 2.0$ $\qquad S_{xy} = \sum (x-\bar{x})(y-\bar{y}) = -20.2$

$S_{yy} = \sum (y-\bar{y})^2 = 12.64$

So,

$$\hat{\beta}_1 = \frac{S_{xy}}{S_{xx}} = -0.5266, \quad \hat{\beta}_0 = \bar{y} - \hat{\beta}_1\bar{x} = 2.0 - (-0.5266)(3.6429) = 3.9184$$

The fitted line is $\hat{y} = 3.92 - 0.53x$, which is graphed in part (a).

(c) Proportion of y variability explained is given by $r^2 = \frac{S_{xy}^2}{S_{xx}S_{yy}} = 0.842$.

12.3 (a) $y' = \frac{1}{y^{1/3}}, \quad x' = x$

(b) $y' = \frac{1}{y}, \quad x' = \frac{1}{1+x}$

12.5 (a) The scatter diagram is shown in the below figure (labeled as (i)).

(b) The scatter diagram of the transformed data, $x' = \log x$ and $y' = \log y$, reveals a more nearly linear relationship – this is illustrated in the figure below (labeled as (ii)).

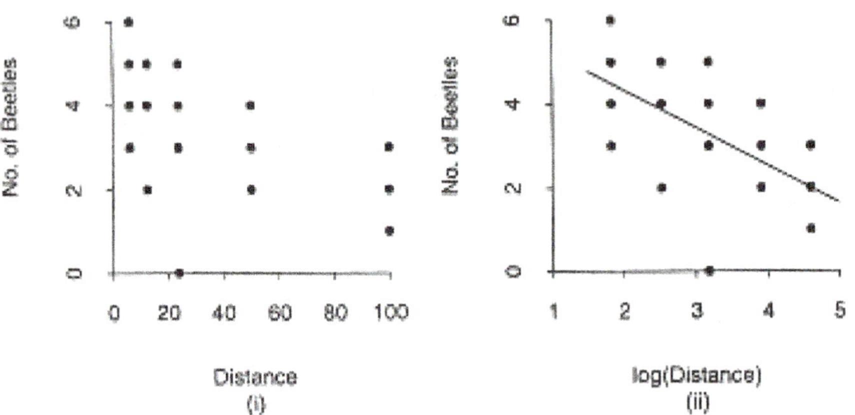

Using Minitab, the original data are entered into columns C1 and C2 – the output is at the top of the next page.

In order to evaluate the mean and sum of squares for log(x), we also calculate the following using Minitab:

```
          STDEV C3
ST.DEV. =        1.0059
          MEAN C3
MEAN    =        3.2107
```

```
NAME C1 'DISTANCE' C2 'BEETLES' C3 'LOGEDIST'
LOGE C1 SET IN C3
REGRESS Y IN C2 ON 1 PREDICTOR IN C3

THE REGRESSION EQUATION IS
BEETLES = 6.14 - 0.899 LOGEDIST

PREDICTOR       COEF        STDEV    T-RATIO        P
CONSTANT      6.1e74       0.9918       6.19    0.000
LOGEDIST     -0.8993       0.2954      -3.04    0.007

S = 1.295        R-SQ = 34.0%      R-SQ(ADJ) = 30.3%

ANALYSIS OF VARIANCE

SOURCE        DF          SS          MS        F        P
REGRESSION     1      15.547      15.547     9.27    0.007
ERROR         18      30.203       1.678
TOTAL         19      45.750
```

(c) Since $S_{xx} = (n-1)(\text{Standard Deviation})^2 = 19(1.0059)^2 = 19.225$ and $t_{0.025} = 2.101$ for d.f. = 18, a 95% confidence interval for β_1 is given by

$$\hat{\beta}_1 \pm 2.101 \frac{s}{\sqrt{S_{xx}}} = -0.8993 \pm 2.101(0.2954) \text{ or } (-1.52, -0.28).$$

(d) At $x = 18$, $\hat{y} = 6.14 - 0.899 \log_e(18) = 3.54$. Since $S_{xx} = 19.225$, a 95% confidence interval is given by

$$3.54 \pm 2.101(1.295)\sqrt{\frac{1}{20} + \frac{(2.8904 - 3.2107)^2}{19.225}} = 3.54 \pm 0.64$$

or (2.90, 4.18).

12.7 When $x_1 = 3$ and $x_2 = -2$, the mean of the response Y is

$$E(Y) = \beta_0 + \beta_1 x_1 + \beta_2 x_2 = -1 - 2(3) + 3(-2) = -13.$$

12.9 (a) Since $t_{0.025} = 2.110$ for d.f. = 48, the 95% confidence interval for β_2 is

$$\hat{\beta}_2 \pm t_{0.025} \times \text{SE}(\hat{\beta}_2) = -1.560 \pm 1.68(0.1862) \text{ or } (-1.873, -1.247)$$

(b) Given that $\alpha = 0.05$ and H_1 is right-sided, the rejection region is $R: T > t_{0.05} = 1.68$ for d.f. = 48. Since the observed value of t is

$$\frac{\hat{\beta}_1 - 0.0125}{\text{SE}(\beta_1)} = \frac{0.0156 - 0.0125}{0.00161} = 1.925,$$ which lies in R, we reject the null hypothesis $H_0: \beta_1 = 0.0125$ in favor of $H_1: \beta_1 > 0.0125$ at $\alpha = 0.05$.

12.11 (a) $\hat{\beta}_0 = 45.3, \ \hat{\beta}_1 = -3.22, \ \hat{\beta}_2 = -0.02066$

(b) $\hat{y} = 45.3 - 3.22x_1 - 0.02066x_2$.

(c) The proportion of y variability explained is $R^2 = 0.795$.

(d) $s^2 = \dfrac{\text{SSE}}{n-2} = \text{MSE} = 11.58$

12.13 (a) Given that $\alpha = 0.05$ and H_1 is two-sided, the rejection region is $R: |T| > t_{0.025} = 2.085$ for d.f. = 20. Since the observed value of t is $\dfrac{\hat{\beta}_1 - 0}{\text{SE}(\beta_1)} = \dfrac{-3.22 - 0}{0.4562} = -7.058$, which lies in R, we reject the null hypothesis $H_0: \beta_1 = 0$ in favor of $H_1: \beta_1 \neq 0$ at $\alpha = 0.05$.

(b) Given that $\alpha = 0.05$ and H_1 is two-sided, the rejection region is $R: |T| > t_{0.025} = 2.085$ for d.f. = 20. Since the observed value of t for $\hat{\beta}_2$ is 2.414, which does lie in R, we reject the null hypothesis $H_0: \beta_2 = 0$ in favor of $H_1: \beta_2 \neq 0$ at $\alpha = 0.05$.

(c) $\hat{y} = 45.3 - 3.22(3.2) - 0.0207(2.0) = 34.955$

(d) A 90% confidence interval for β_0, which does not include 0, is given by $\hat{\beta}_0 \pm t_{0.05} \times \text{SE}(\hat{\beta}_0) = 45.3 \pm 1.771(2.345)$ or $(41.147, 49.453)$.

12.15 (a) The scatter diagram of y versus $x' = \log_{10} x$ is shown below:

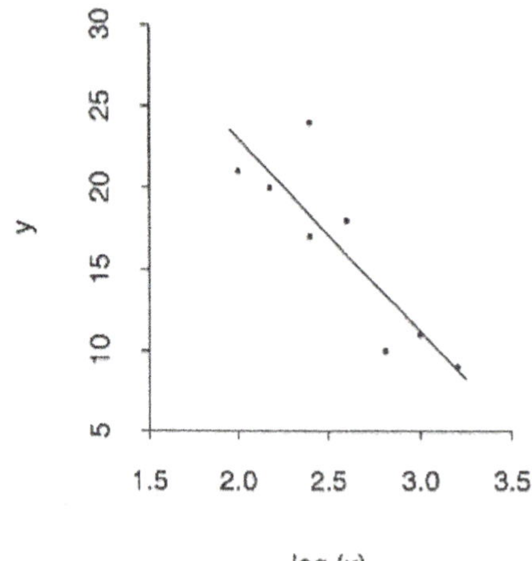

(b) The Minitab output is given below. The fitted line is $\hat{y} = 46.55 - 11.77 \log_{10} x$ (which is shown in part (a)).

```
The regression equation is
y = 46.6 - 11.8 logt(x)

Predictor      Coef     StDev        T        P
Constant     46.550     7.242     6.43    0.001
logt(x)     -11.772     2.783    -4.23    0.005


  S = 3.031          R-Sq = 74.9%

  Analysis of Variance

  Source          DF        SS        MS       F      P
  Regression       1    164.39    164.39   17.90  0.005
  Residual Error   6     55.11      9.19
  Total            7    219.50
```

(c) From the Minitab output in part (b), $\frac{s}{\sqrt{S_{x'x'}}} = 2.788$. You could also obtain the same result from direct calculation using $S_{x'x'} = 1.1862$. A 90% confidence interval for β_1 is given by

$$\hat{\beta}_1 \pm 1.943 \frac{s}{\sqrt{S_{x'x'}}} = -11.77 \pm 1.943(2.788) \text{ or } (-17.18, -6.36).$$

(d) At $x = 300$, $\hat{y} = 46.55 - 11.772(\log_{10}(300)) = 17.39$. Since $\bar{x}' = 2.5739$ and $S_{x'x'} = 1.1862$, a 95% confidence interval for the expected y-value at $x = 300$ is given by

$$\hat{y} \pm t_{0.025}\, s\sqrt{\frac{1}{n} + \frac{\left((x')^* - \bar{x}'\right)^2}{S_{x'x'}}}$$

$$= 17.39 \pm 2.447(3.031)\sqrt{\frac{1}{8} + \frac{(2.4471 - 2.5739)^2}{1.1862}}$$

$$= 17.39 \pm 2.76$$

or (14.6, 20.15).

12.17 (a) & (b) The scatter diagram of y versus x shown in the figure (part (i)) below reveals a relation along a curve, and that of $y' = \log_{10}(y)$ versus x shown in the figure (part (ii)) looks like a straight line relation.

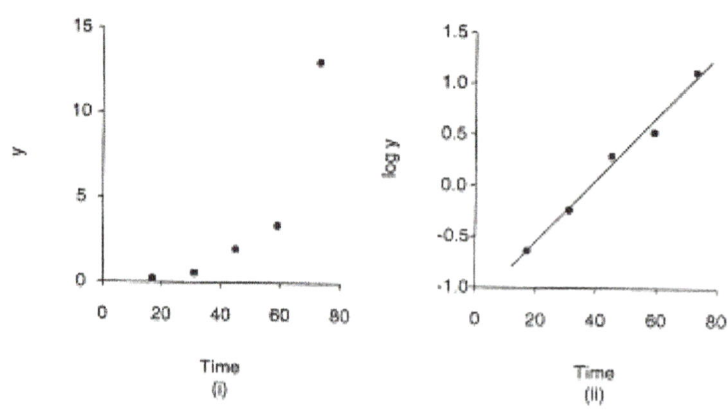

(c) The fitted line $\widehat{\log_{10}(y)} = -1.16 + 0.0305x$ is shown in part (ii) of the figure in part (a). The Minitab output is shown below:

```
    NAME C1 'TIME' C2 'Y' C3 'LOG(Y)'
    LOGT C2 SET C3
    REGRESS Y IN C3 ON 1 PREDICTOR IN C1

THE REGRESSION EQUATION IS
LOG(Y) = - 1.16 + 0.0305 TIME

PREDICTOR        COEF       STDEV     T-RATIO       P
CONSTANT     -1.15984     0.09468      -12.25   0.001
TIME         0.030513    0.001926       15.84   0.001

S = 0.08526       R-SQ = 98.8%      R-SQ(ADJ) = 98.4%

ANALYSIS OF VARIANCE

SOURCE       DF       SS        MS         F       P
REGRESSION    1   1.8249    1.8249    251.05   0.001
ERROR         3   0.0218    0.0073
TOTAL         4   1.8467
```

12.19 (a) A 90% confidence interval for β_1 is given by

$\hat{\beta}_1 \pm t_{0.05} \times \text{SE}(\hat{\beta}_1) = 0.2869 \pm 1.752(0.0606)$ or $(0.1807, 0.3931)$.

(b) Given that $\alpha = 0.05$ and H_1 is two-sided, the rejection region is $R: |T| > t_{0.025} = 2.101$ for d.f. = 15. Since the observed value of t for $\hat{\beta}_2$ is

$\frac{\hat{\beta}_2 - 0.5}{\text{SE}(\hat{\beta}_2)} = \frac{0.6532 - 0.5}{0.0632} = 2.424$, which lies in R, we reject the null hypothesis $H_0 : \beta_2 = 0.5$ in favor of $H_1 : \beta_2 \neq 0.5$ at $\alpha = 0.05$.

12.21 (a) We plot the residual versus the predicted value $\hat{y}$ and the time order, respectively, in parts (i) and (ii) of the below figure.

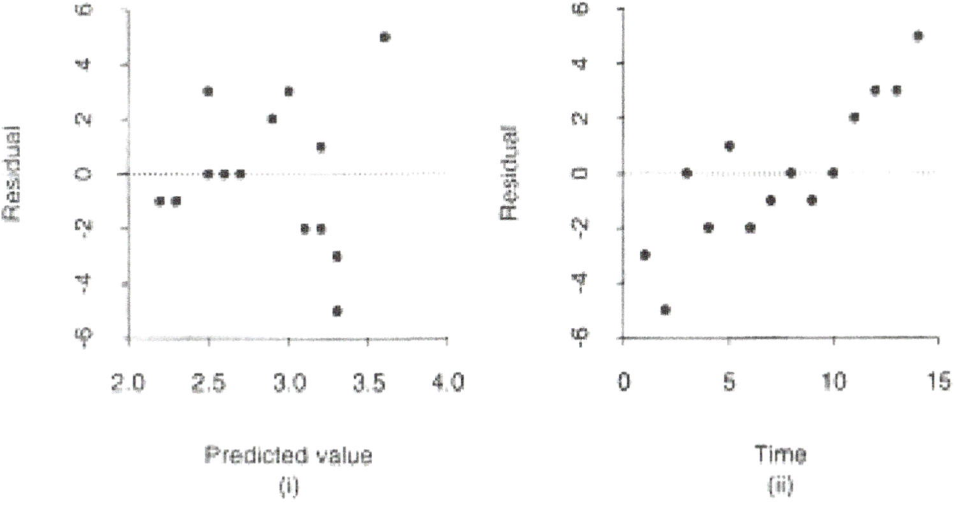

(b) The plot (i) does not seem to signify any appreciable violation of the assumptions. The plot (ii) of residual versus time order, however, exhibits a distinct pattern. The residuals tend to steadily increase in time. This indicates a possible violation of the independence assumption.

12.23 Looking at the residuals in time order shown in the table, a distinct pattern is apparent. The residuals are all positive for the first seven years, though in steady decline, and then there is a discernible change in year 8 that persists for all subsequent years. This pattern casts serious doubt on the independence assumption. Violation of independence is frequent in time series data such as these.

12.25 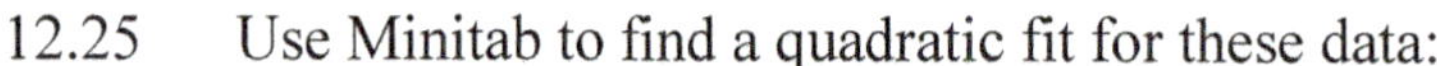
Use Minitab to find a quadratic fit for these data:

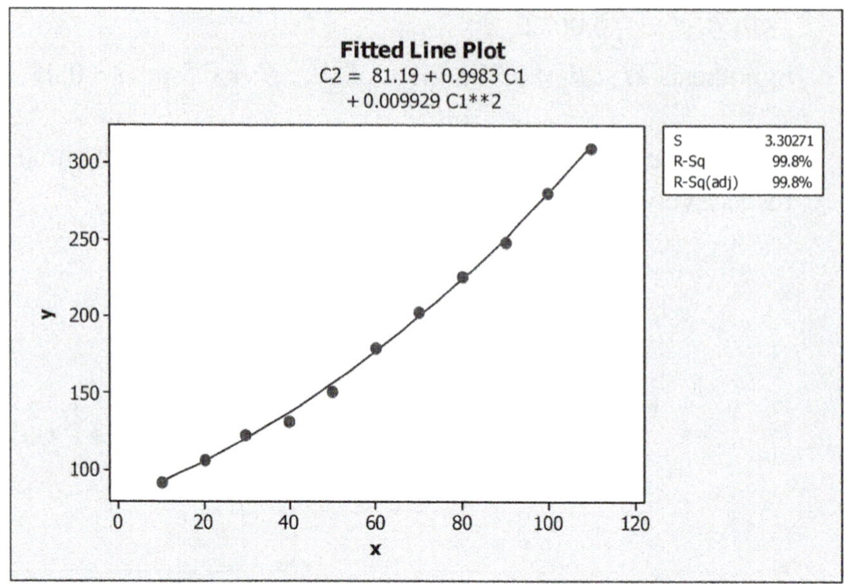


```
The regression equation is
     C2 = 81.19 + 0.9983 C1 + 0.009929 C1**2

     S = 3.30271   R-Sq = 99.8%   R-Sq(adj) = 99.8%

     Analysis of Variance

     Source      DF        SS        MS        F       P
     Regression   2   53589.8   26794.9  2456.47   0.000
     Error        8      87.3      10.9
     Total       10   53677.1

     Sequential Analysis of Variance

     Source     DF        SS       F      P
     Linear      1   52744.0  508.73  0.000
     Quadratic   1     845.8   77.54  0.000
```

(b) The proportion of y variability explained is $R^2 = 0.998$, or 99.8%.

(c) Given that $\alpha = 0.05$ and H_1 is left-sided, the rejection region is $R: T < -t_{0.05} = -1.860$ for d.f. = 8. Since the observed value of t is $\dfrac{\hat{\beta}_1 - 0}{\text{SE}(\beta_1)} = \dfrac{0.9983}{0.1389} = 7.19$, we reject reject the null hypothesis $H_0 : \beta_1 = 0$ in favor of $H_1 : \beta_1 < 0$ at $\alpha = 0.05$.

12.27 (a) The Minitab output is as follows:

```
SQRT C2 SET IN C3
REGRESS Y IN C3 ON 1 PREDICTOR IN C1

THE REGRESSION EQUATION IS
C3 = - 0.167 + 0.237 C1

PREDICTOR        COEF        STDEV     T-RATIO        P
CONSTANT      -0.1665       0.9323       -0.18    0.863
C1            0.23703      0.02383        9.95    0.000

S = 0.9563        R-SQ = 92.5%      R-SQ(ADJ) = 91.6%

ANALYSIS OF VARIANCE

SOURCE       DF          SS          MS          F         P
REGRESSION    1      90.456      90.456      98.91     0.000
ERROR         8       7.316       0.915
TOTAL         9      97.773
```

From the output, note that the fitted line is $\hat{y}' = -0.167 + 0.237x$ and $r^2 = 0.925$. The constant term could be dropped and the model re-fit.

(b) Since $t_{0.025} = 2.306$ for d.f. = 8, a 95% confidence interval for β_1 is given by $\hat{\beta}_1 \pm 2.306\frac{s}{\sqrt{S_{xx}}} = 0.23703 \pm 2.306(0.02383)$ or $(0.182, 0.292)$.

(c) Since $\bar{x} = 37$ and $S_{xx} = 1610$, a 95% confidence interval (for $x = 45$) is given by

$$[-0.167 + 0.237(45)] \pm 2.306(0.9563)\sqrt{\frac{1}{10} + \frac{(45-37)^2}{1610}} \text{ or } (9.67, 11.32).$$

12.29 (a) $\hat{y} = 50.4 + 0.1907x_2$ and $r^2 = 0.03$. The Minitab output is shown below:

```
NAME C1 'X1' C2 'X2' C3 'X3' C4 'Y'
REGRESS Y IN C4 ON 1 PREDICTOR IN C2

THE REGRESSION EQUATION IS
Y = 50.4 + 0.191 X2

PREDICTOR      COEF     STDEV   T-RATIO        P
CONSTANT      50.40     17.57      2.87    0.010
X2           0.1907    0.2561      0.74    0.466

S = 19.95          R-SQ = 3.0%  R-SQ(ADJ) = 0.0%

ANALYSIS OF VARIANCE

SOURCE       DF        SS        MS      F       P
REGRESSION    1     220.8     220.8   0.55   0.466
ERROR        18    7164.2     398.0
TOTAL        19    7385.0
```

(b) $\hat{y} = -92.32 + 0.583x_1 - 0.1494x_2 + 35.07x_3$ and $R^2 = 0.586$. The Minitab output is shown below:

```
REGRESS Y IN C4 ON 3 PREDICTORS IN C1 C2 C3

THE REGRESSION EQUATION IS
Y = - 92.3 + 0.583 X1 - 0.149 X2 + 35.1 X3

PREDICTOR      COEF      STDEV    T-RATIO        P
CONSTANT     -92.32      33.22      -2.78    0.013
X1           0.5830     0.3206       1.82    0.088
X2          -0.1494     0.1923      -0.78    0.449
X3            35.07      11.25       3.12    0.007
```

```
S = 13.83     R-SQ = 58.6%      R-SQ(ADJ) = 50.8%

ANALYSIS OF VARIANCE

SOURCE        DF      SS       MS       F     P
REGRESSION     3  4325.7   1441.9    7.54 0.002
ERROR         16  3059.2    191.2
TOTAL         19  7385.0
SOURCE        DF     SEQ SS
X1             1     2466.6
X2             1        1.4
X3             1     1857.8
```

(c) Even three variables do not predict well. In fact, the GPA (x_3) could predict almost as well by itself. We summarize the results in the following table:

Predictor	x_3	x_3 and x_1	x_3, x_1, and x_2
R^2	0.495	0.570	0.586

12.31 The design matrix ***X*** of the model $Y = \beta_0 + \beta_1 x_1 + \beta_2 x_2 + e$ is

$$X = \begin{bmatrix} 1 & 1 & 9 \\ 1 & 2 & 8 \\ 1 & 3 & 27 \\ 1 & 3 & 32 \\ 1 & 4 & 34 \\ 1 & 5 & 48 \\ 1 & 5 & 54 \\ 1 & 6 & 44 \\ 1 & 6 & 80 \\ 1 & 7 & 122 \end{bmatrix}$$

12.33 Using Minitab, the data for gender, initial and final number of sit-ups are entered into columns C1, C8, and C9, respectively. The fitted equation is

final number = 9.999 + 0.155 (gender) + 0.9015 (initial number)

With the high p-value of 0.899 shown in the output (below), we cannot reject the hypothesis that the coefficient of predictor 'gender' is zero. In fact, a simple linear fit with initial number of sit-ups as predictor has $r^2 = 0.715$.

```
      NAME C1 'GENDER' C8 'PRETEST' C9 'POSTTEST'
      REGRESS Y IN C9 ON 2 PREDICTORS IN C1 AND C8

THE REGRESSION EQUATION IS
POSTTEST = 10.0 + 0.15 GENDER + 0.902 PRETEST

PREDICTOR        COEF       STDEV    T-RATIO        P
CONSTANT        9.999       3.640       2.75    0.007
GENDER          0.155       1.211       0.13    0.899
PRETEST       0.90150     0.06710      13.43    0.000

S = 5.211        R-SQ = 71.5%      R-SQ(ADJ) = 70.8%

ANALYSIS OF VARIANCE

SOURCE       DF        SS       MS          F        P
REGRESSION    2    5313.4   2656.7      97.82    0.000
ERROR        78    2118.4     27.2
TOTAL        80    7431.8

SOURCE       DF    SEQ SS
GENDER        1     411.6
PRETEST       1    4901.7
```

Chapter 13

ANALYSIS OF CATEGORICAL DATA

13.1 The null hypothesis is $H_0: p_1 = p_2 = ... = p_6 = \frac{1}{6}$ and the alternative hypothesis H_1 is that at least two proportions are different. Under H_0, the expected frequency of each cell is $320\left(\frac{1}{6}\right)$. The χ^2 statistic for goodness-of-fit is calculated as follows:

Face Number	1	2	3	4	5	6	Total
Observed Frequency (O)	39	63	56	67	57	38	320
Expected Frequency (E)	$\frac{320}{6}$	$\frac{320}{6}$	$\frac{320}{6}$	$\frac{320}{6}$	$\frac{320}{6}$	$\frac{320}{6}$	320
$\frac{(O-E)^2}{E}$	3.852	1.752	0.133	3.502	0.252	4.408	$13.90 = \chi^2$

We take $\alpha = 0.05$. For d.f. = 5, we find that $\chi^2_{0.05} = 11.07$, so that the rejection region is $R: \chi^2 \geq 11.07$. Since the observed value of $\chi^2 = 13.90$ lies in R, we reject H_0 at $\alpha = 0.05$. As such, the model of a fair die is contradicted.

13.3 The null hypothesis is $H_0: p_1 = 0.4,\ p_2 = 0.4,\ p_3 = 0.1,\ p_4 = 0.1$ and the alternative hypothesis H_1 is that at least one of these proportions is not correct. Under H_0, multiplying these probabilities by $n = 100$, the expected frequencies are found to be 40, 40, 10, and 10. The χ^2 statistic for goodness-of-fit is calculated as follows:

Blood Type	O	A	B	AB	Total
Observed Frequency (O)	40	44	10	6	100
Expected Frequency (E)	40	40	10	10	100
$\frac{(O-E)^2}{E}$	0.00	0.40	0.00	1.60	$2.00 = \chi^2$

We take $\alpha = 0.05$. For d.f. = 3, we find that $\chi^2_{0.05} = 7.81$, so that the rejection region is $R: \chi^2 \geq 7.81$. Since the observed value of $\chi^2 = 2.00$ does not lie in R, we do not reject H_0 at $\alpha = 0.05$.

13.5 Let p_1, p_2, p_3, and p_4 denote the population proportions of walnuts, hazelnuts, almonds, and pistachios, respectively. The null hypothesis is $H_0: p_1 = 0.45,\ p_2 = 0.20,\ p_3 = 0.20,\ p_4 = 0.15$ and the alternative hypothesis H_1 is that at least one of these proportions is not correct. Under H_0, multiplying these probabilities by $n = 240$, the expected frequencies are found to be 108, 48, 48, and 36. The χ^2 statistic for goodness-of-fit is calculated as follows:

Nut Type	Walnuts	Hazelnuts	Almonds	Pistachios	Total
Observed Frequency (O)	95	70	33	42	240
Expected Frequency (E)	108	48	48	36	240
$\frac{(O-E)^2}{E}$	1.565	10.083	4.688	1.000	$17.336 = \chi^2$

We take $\alpha = 0.025$. For d.f. = 3, we find that $\chi^2_{0.025} = 9.35$, so that the rejection region is $R: \chi^2 \geq 9.35$. Since the observed value of $\chi^2 = 17.336$ lies in R, we reject H_0 at $\alpha = 0.025$. We conclude that there is strong evidence of mislabeling.

13.7 For the particular geographical region, let $p_1, p_2, \ldots, p_6$ denote the population proportions of accidental deaths due to motor vehicle, poison, falls, choking, drowning, and other reasons, respectively. The null hypothesis is $H_0: p_1 = 0.405,\ p_2 = 0.185,\ p_3 = 0.157,\ p_4 = 0.041,\ p_5 = 0.032,\ p_6 = 0.180$ and the alternative hypothesis H_1 is that at least one of these proportions is not correct.

Under H_0, multiplying these probabilities by $n = 908$, the expected frequencies are found to be 367.74, 167.98, 142.56, 37.23, 29.06 and 163.44. The χ^2 statistic for goodness-of-fit is calculated as follows:

Reason	Motor Vehicle	Poison	Falls	Choking	Drowning	Other	Total
Observed Frequency (O)	356	207	125	33	26	161	908
Expected Frequency (E)	367.74	167.98	142.56	37.23	29.06	163.44	908
$\frac{(O-E)^2}{E}$	0.375	9.064	2.163	0.481	0.322	0.036	12.441 $= \chi^2$

We take $\alpha = 0.05$. For d.f. = 5, we find that $\chi^2_{0.05} = 11.07$, so that the rejection region is $R: \chi^2 \geq 11.07$. Since the observed value of χ^2 = 12.441 lies in R, we reject H_0 at $\alpha = 0.05$. As such, we conclude that the pattern is significantly different – the differences in categories 2 and 3 are conspicuous from the large values of $\frac{(O-E)^2}{E}$.

13.9 (a) From the binomial table for $n = 3$ and $p = 0.4$, we find that

$$P[X = 0] = 0.216$$
$$P[X = 1] = 0.648 - 0.216 = 0.432$$
$$P[X = 2] = 0.936 - 0.648 = 0.288$$
$$P[X = 3] = 1.000 - 0.936 = 0.064$$

(b) Let p_0, p_1, p_2, and p_3 denote the probabilities of the four categories: 0, 1, 2 and 3 males in litter, respectively. The null hypothesis is $H_0: p_0 = 0.216,\ p_1 = 0.432,\ p_2 = 0.288,\ p_3 = 0.064$ and the alternative hypothesis H_1 is that at least one of these proportions is not correct. Under H_0, multiplying these probabilities by $n = 80$, the expected frequencies are found to be 17.28, 34.56, 23.04, and 5.12. The χ^2 statistic for goodness-of-fit is calculated as follows:

Number of Males	0	1	2	3	Total
Observed Frequency (O)	19	32	22	7	80
Expected Frequency (E)	17.28	34.56	23.04	5.12	80
$\frac{(O-E)^2}{E}$	0.171	0.190	0.047	0.690	1.098 $= \chi^2$

We take $\alpha = 0.05$. For d.f. = 2, we find that $\chi^2_{0.05} = 5.99$, so that the rejection region is $R: \chi^2 \geq 5.99$. Since the observed value of $\chi^2 = 1.098$ does not lie in R, we do not reject H_0 at $\alpha = 0.05$. As such, we conclude that the binomial model is not contradicted.

13.11 (a) We summarize the information in the table below:

	Open all mail	Don't open all mail	Total
Males	414	386	800
Females	532	368	900
Total	946	754	1700

(b) Let p_1 and p_2 be the probabilities that a person opens all of his/her mail for males and females, respectively.
The null hypothesis of homogeneity is $H_0: p_1 = p_2$.

(c) We take $\alpha = 0.05$. For d.f. = 1, we find that $\chi^2_{0.05} = 3.84$, so that the rejection region is $R: \chi^2 \geq 3.84$. The χ^2 statistic for homogeneity is calculated as follows:

Expected Values:

	Open all mail	Don't open all mail	Total
Males	445.176	354.824	800
Females	500.824	399.176	900
Total	946	754	1700

$\dfrac{(O-E)^2}{E}$:

	Open all mail	Don't open all mail
Males	2.183	2.739
Females	1.941	2.435

Then, $\chi^2 = 2.183 + 2.739 + 1.941 + 2.435 = 9.298$, which lies in R. So, we reject H_0 (of equal probabilities of opening all mail) at $\alpha = 0.05$.
Furthermore, the p-value is about 0.002.

(d) The largest contribution to χ^2 comes from the mail – don't open cell, where the observed count is higher than expected. The female – don't open count is lower than expected.

13.13 Denote by p_{A1}, p_{A2}, p_{A3}, and p_{A4} the probabilities of response in the categories 'none', 'slight', 'moderate', and 'severe', respectively, under the use of Brand A pills. Similarly, use p_{B1}, p_{B2}, p_{B3}, and p_{B4} for Brand B. We are to test the null hypothesis of homogeneity $H_0 : p_{Aj} = p_{Bj}$ $(j = 1,2,3,4)$. The χ^2 statistic for homogeneity is calculated as follows:

```
Expected counts are printed below observed counts

           C1       C2       C3       C4    Total
    1      18       17        6        4       45
        14.50    15.50    10.00     5.00

    2      11       14       14        6       45
        14.50    15.50    10.00     5.00

Total      29       31       20       10       90

ChiSq =  0.845 +  0.145 +  1.600 +  0.200 +
         0.845 +  0.145 +  1.600 +  0.200 = 5.580
df = 3
```

We take $\alpha = 0.05$. For d.f. = 3, we find that $\chi^2_{0.05} = 7.81$, so that the rejection region is $R: \chi^2 \geq 7.81$. Since the observed $\chi^2 = 5.580$ (from the Minitab output) does not lie in *R*, we do not reject H_0 at $\alpha = 0.05$. As such, we fail to conclude that the two pills are significantly different in quality.

13.15 We form the following table of observed values:

Observed	Over-reported	Not over-reported	Total
Bottom Signature	26	7	33
No Signature	21	12	33
TOTAL	47	19	66

Let

P_{OB} and P_{ON} be the probability of over reporting for groups where the signature is on the bottom or not required, respectively,
and
P_{UB} and P_{UN} be the probability of under reporting for groups where the signature is on the bottom or not required, respectively.

The null hypothesis for a test of homogeneity is: H_0: $P_{OB} = P_{ON} = P_{UB} = P_{UN}$.

Use $\alpha = 0.05$. For df = 1, $\chi^2_{0.05} = 3.841$, so the rejection region is R: $\chi^2 \geq 3.841$. We calculate the test statistic as follows:

Expected Values	Over-reporting	Not over reporting
Bottom Signature	$\frac{33\times 47}{66} = 23.5$	$\frac{33\times 19}{66} = 9.5$
No Signature	$\frac{33\times 47}{66} = 23.5$	$\frac{33\times 19}{66} = 9.5$

$\frac{(O-E)^2}{E}$	Over-reporting	Not over reporting
Bottom Signature	0.267	0.658
No Signature	0.267	0.658

So, $\chi^2 =$ Sum of the tabulated values $= 1.85$.
This value does not lie in the rejection region. So, we do not reject the null hypothesis at this significance level.

(b) The results suggest that the data does not provide strong enough evidence to conclude that the probabilities are different.

13.17 (a) We form the following table of observed values:

Observed	Center thickness $\geq$ 1.9mm	Center thickness < 1.9mm	Total
Passed impact test	49	13	62
Failed impact test	12	18	30
TOTAL	61	31	92

The null hypothesis for a test of homogeneity is: H_0: No difference between passing and failing the impact test

Use $\alpha = 0.05$. For df = 1, $\chi^2_{0.05} = 3.841$, so the rejection region is R: $\chi^2 \geq 3.841$.
We calculate the test statistic as follows:

Expected Values	Center thickness $\geq$ 1.9mm	Center thickness < 1.9mm
Passed impact test	$\frac{62\times 61}{92} = 41.11$	$\frac{62\times 31}{92} = 20.89$
Failed impact test	$\frac{30\times 61}{92} = 19.89$	$\frac{30\times 31}{92} = 10.11$

$\frac{(O-E)^2}{E}$	Center thickness $\geq$ 1.9mm	Center thickness < 1.9mm
Passed impact test	1.51	2.98
Failed impact test	3.13	6.16

So, $\chi^2 =$ Sum of the tabulated values $= 13.78$.
This value lies in the rejection region. So, we reject the null hypothesis at this significance level.

(b) The number of lenses whose centers is thicker than 1.9mm and which pass the impact test constitute more than half of the observed value and so, it likely what is making the probabilities become unequal.

13.19 (a) The calculations are identical.
(b) The Minitab output is as follows:

```
Chi-Square Test


Expected counts are printed below observed counts

          Open Dont Ope    Total
    1      414      386      800
           445.18   354.82

    2      532      368      900
           500.82   399.18

Total      946      754     1700

  Chi-Sq =  2.183 +  2.739 +
            1.941 +  2.435 = 9.298
DF = 1, P-Value = 0.002
```

We take $\alpha = 0.05$. For d.f. = 1, we find that $\chi^2_{0.05} = 3.84$, so that the rejection region is $R: \chi^2 \geq 3.84$. Since the observed $\chi^2 = 9.298$ (from the Minitab output) lies in *R*, we reject H_0 at $\alpha = 0.05$. Furthermore, the associated *p*-value is 0.002. As such, the proportions of males and females who open all of their mail are significantly different.

(c) The Minitab output is as follows:

```
Expected counts are printed below observed counts

    C1        C2        C3       Total
    1         38        15         7         60
              22.50     23.10     14.40

    2         22        32        16         70
              26.25     26.95     16.80

    3         15        30        25         70
              26.25     26.95     16.80

Total         75        77        48        200
   ChiSq =  10.678 +  2.840 +  3.803 +
             0.688 +  0.946 +  0.038 +
             4.821 +  0.345 +  4.002 = 28.162
df = 4
```

We take $\alpha = 0.01$. For d.f. = 4, we find that $\chi^2_{0.01} = 13.28$, so that the rejection region is $R: \chi^2 \geq 13.28$. Since the observed $\chi^2 = 28.162$ (from the Minitab output) lies in *R*, we reject H_0 at $\alpha = 0.01$. Comparing observed and expected frequencies, we see that the bone loss is higher in the control group than the activity group.

13.21 We test the null hypothesis that the pattern of appeals decision and the type of representation are independent. The χ^2 statistic is calculated using Minitab as follows:

```
Expected counts are printed below observed counts

          C1       C2       C3    Total
    1     59      108       17      184
          74.18    98.32    11.50

    2     70       63        3      136
          54.83    72.68     8.50

Total    129      171       20      320

 ChiSq =  3.105 +  0.952 +  2.630 +
          4.200 +  1.288 +  3.559 = 15.734
df = 2
```

We take $\alpha = 0.05$. For d.f. = 2, we find that $\chi^2_{0.05} = 5.99$, so that the rejection region is $R: \chi^2 \geq 5.99$. Since the observed $\chi^2 = 15.734$ (from the Minitab output) lies in *R*, we reject H_0 at $\alpha = 0.05$. We conclude that the patterns of appeals decision are significantly different between the two types of representation.

13.23 We test the null hypothesis of independence between union membership and attitude toward spending on social welfare. The χ^2 statistic is calculated using Minitab as follows:

```
Expected counts are printed below observed counts

          C1       C2       C3    Total
    1    112       36       28      176
          86.24    45.76    44.00

    2     84       68       72      224
```

```
            109.76     58.24     56.00

Total       196        104       100       400

ChiSq =  7.695 +  2.082 +  5.818 +
         6.046 +  1.636 +  4.571 = 27.847

df = 2
```

We take $\alpha = 0.01$. For d.f. = 2, we find that $\chi^2_{0.01} = 9.21$, so that the rejection region is $R: \chi^2 \geq 9.21$. Since the observed $\chi^2 = 27.847$ (from the Minitab output) lies in *R*, we reject H_0 at $\alpha = 0.01$. We conclude that attitudes and union membership are dependent. There are significant differences between the attitudes of the union and non-union groups.

13.25 The null hypothesis is that group and stopping response are independent. The cell probabilities are the product of the marginal probabilities. The χ^2 statistic is calculated using Minitab as follows:

```
Expected counts are printed below observed counts

          C1        C2     Total
1          9         9        18
        6.55     11.45

2          8        12        20
        7.27     12.73


          3          3        14        17
                  6.18     10.82

     Total          20        35        55

     ChiSq =   0.920 +   0.526 +
               0.073 +   0.042 +
               1.638 +   0.936 = 4.134
df = 2
```

We take $\alpha = 0.05$. For d.f. = 2, we find that $\chi^2_{0.05} = 5.99$, so that the rejection region is $R: \chi^2 \geq 5.99$. Since the observed $\chi^2 = 4.134$ (from the Minitab output) does not lie in *R*, we do not reject H_0 at $\alpha = 0.05$. We conclude that, under independence, the groups are not significantly different in their response.

13.27 (a) They are identical
(b) Doing this on Minitab yields the same output.

13.29 Denote by $p_0, p_1, \ldots, p_9$ the probabilities of the integers 0, 1, 2, … , 9, respectively. The null hypothesis that all ten integers are equally likely is formalized by $H_0: p_0 = p_1 = \ldots = p_9 = \frac{1}{10}$ and the alternative hypothesis H_1 is that at least two of these proportions are different. Under H_0, the expected frequency of each cell is $500\left(\frac{1}{10}\right) = 50$. The χ^2 statistic for goodness-of-fit is calculated as follows:

Integer	0	1	2	3	4	5	6	7	8	9	Total
Observed Frequency (O)	41	58	51	61	39	56	45	35	62	52	500
Expected Frequency (E)	50	50	50	50	50	50	50	50	50	50	500
$\frac{(O-E)^2}{E}$	1.62	1.28	0.02	2.42	2.42	0.72	0.50	4.50	2.88	0.08	16.44 $= \chi^2$

We take $\alpha = 0.05$. For d.f. = 9, we find that $\chi^2_{0.05} = 16.92$, so that the rejection region is $R: \chi^2 \geq 16.92$. Since the observed value of $\chi^2 = 16.44$ does not lie in R, we do not reject H_0 at $\alpha = 0.05$. As such, the data do not demonstrate any bias.

13.31

(a) Let $p_{1i}, p_{2i}, p_{3i}, p_{4i}$ denote the population proportions from 1, 2, 3, and 4 respectively, and $i = 1$ stands for papers delivered and $i = 2$ stands for papers remaining.

(b) We test the null hypothesis of homogeneity: $H_0: p_{1i} = p_{2i} = p_{3i} = p_{4i}$ $(i = 1, 2)$. Take $\alpha = 0.05$. For d.f. = 3, $\chi^2_{0.05} = 7.81$ and so the rejection region is $R: \chi^2 \geq 7.81$.

Observed Frequencies O:

	Papers Delivered	**Papers Remaining**	
Site 1	50	17	67
Site 2	47	12	59
Site 3	48	7	55
Site 4	50	21	71
TOTAL	195	57	252

Expected Values E:

	Papers Delivered	**Papers Remaining**
Site 1	51.8	15.2
Site 2	45.7	13.3
Site 3	42.6	12.4
Site 4	54.9	16.1

$\frac{(O-E)^2}{E}$:

	Papers Delivered	Papers Remaining
Site 1	0.063	0.213
Site 2	0.037	0.127
Site 3	0.685	2.351
Site 4	0.437	1.491

Take $\alpha = 0.05$. For d.f. = 3, $\chi^2_{0.05} = 7.81$ and so the rejection region is $R : \chi^2 \geq 7.81$. Thus, the test statistic is $\chi^2 = 5.404$. Since this does not lie in *R*, we do not reject H_0 at $\alpha = 0.05$.

(c) A 95% confidence interval for p_i is given by $\hat{p}_i \pm 1.96\sqrt{\frac{\hat{p}_i\hat{q}_i}{n}}$. We calculate these intervals for each site, as follows:

Site 1: $\hat{p}_1 \pm 1.96\sqrt{\frac{\hat{p}_1\hat{q}_1}{n}} = \frac{33}{50} \pm 1.96\sqrt{\frac{\frac{33}{50}\times\frac{17}{50}}{50}}$

Site 2: $\hat{p}_2 \pm 1.96\sqrt{\frac{\hat{p}_2\hat{q}_2}{n}} = \frac{35}{47} \pm 1.96\sqrt{\frac{\frac{35}{47}\times\frac{12}{47}}{47}}$

Site 3: $\hat{p}_3 \pm 1.96\sqrt{\frac{\hat{p}_3\hat{q}_3}{n}} = \frac{41}{48} \pm 1.96\sqrt{\frac{\frac{41}{48}\times\frac{4}{48}}{48}}$

Site 4: $\hat{p}_4 \pm 1.96\sqrt{\frac{\hat{p}_4\hat{q}_4}{n}} = \frac{29}{50} \pm 1.96\sqrt{\frac{\frac{29}{50}\times\frac{21}{50}}{50}}$

13.33 Denote by p_1 and p_2 the population proportion of persons having hepatitis in the two groups 'vaccinated' and 'not vaccinated', respectively. We are to test $H_0 : p_1 = p_2$ versus $H_1 : p_1 \neq p_2$. The χ^2 statistic for homogeneity is calculated using Minitab as follows:

```
Expected counts are printed below observed counts

           C1        C2     Total
   1       11       538       549
        41.06    507.94

   2       70       464       534
        39.94    494.06

Total      81      1002      1083

 ChiSq = 22.008 +  1.779 +
```

```
                    22.626 +   1.829 = 48.242
df = 1
```

We take $\alpha = 0.05$. For d.f. = 1, we find that $\chi^2_{0.05} = 3.84$, so that the rejection region is $R: \chi^2 \geq 3.84$. Since the observed value of $\chi^2 = 48.242$ lies in *R*, we reject H_0 at $\alpha = 0.05$. This is very strong evidence in support of H_1.

13.35 (a) The two response categories are 'free of pain' and 'not free of pain'. The frequencies of the latter category are obtained by subtracting those of the first from the corresponding 'number of patients assigned'. The 4×2 contingency table is presented here, along with the calculations (from Minitab) for the χ^2 statistic.

```
Expected counts are printed below observed counts

          Free    Not free Total
  1         23        30      53
         27.45     26.55

  2         30        17      47

         24.34     22.66

  3         19        32      51
         26.42     24.58

  4         29        15      44
         22.79     21.21

Total      101        94     195

Chi-Sq =  0.722 +  0.776 +
          1.314 +  1.412 +
          2.082 +  2.237 +
          1.692 +  1.818 = 12.053
DF = 3, P-Value = 0.007
```

We take $\alpha = 0.05$. For d.f. = 3, we find that $\chi^2_{0.05} = 7.81$, so that the rejection region is $R: \chi^2 \geq 7.81$. Since the observed value of $\chi^2 = 12.053$ lies in *R*, we reject H_0 at $\alpha = 0.05$. We conclude that there are significant differences in the effectiveness of the four drugs.

(b) A 90 % confidence interval for a population proportion *p* is given by

$$\hat{p} \pm 1.645\sqrt{\frac{\hat{p}\hat{q}}{n}}.$$

We calculate such an interval for each of the four proportions, as follows:

For Drug 1: $\hat{p}_1 = \frac{23}{53} = 0.434, \quad 0.434 \pm 1.645\sqrt{\frac{(0.434)(0.566)}{53}}$

$= 0.434 \pm 0.112 \quad \text{or} \quad (0.32, 0.55)$

For Drug 2: $\hat{p}_2 = \frac{30}{47} = 0.638, \ 0.638 \pm 1.645\sqrt{\frac{(0.638)(0.362)}{47}}$

$= 0.638 \pm 0.115$ or $(0.52, 0.75)$

For Drug 3: $\hat{p}_3 = \frac{19}{51} = 0.373, \ 0.373 \pm 1.645\sqrt{\frac{(0.373)(0.627)}{51}}$

$= 0.373 \pm 0.111$ or $(0.26, 0.48)$

For Drug 4: $\hat{p}_4 = \frac{29}{44} = 0.659, \ 0.659 \pm 1.645\sqrt{\frac{(0.659)(0.341)}{44}}$

$= 0.659 \pm 0.118$ or $(0.54, 0.78)$

13.37 (a) Test the null hypothesis that each cell probability is the product of the corresponding pairs of marginal probabilities.

The following is a table of the expected values:

	Passed Optical	Failed Optical
Passed Impact	85	34
Failed Impact	25	10

$\frac{(O-E)^2}{E}$	Passed Optical	Failed Optical
Passed Impact	0	0
Failed Impact	0	0

Observe that the test statistic equals $\chi^2 = 0$, so we do not reject the null hypothesis in a big way, at any significance level.

(b) No, but it does suggest that the data is very suspicious.

13.39 We test the null hypothesis that the duration of marriage is independent of the period of acquaintance before marriage. The χ^2 statistic is calculated (using Minitab) as follows:

```
Expected counts are printed below observed counts
          C1        C2    Total
    1     11         8       19
          10.27      8.73


            2        28        24        52
                     28.11     23.89

            3        21        19        40
                     21.62     18.38

      Total          60        51       111

         ChiSq =  0.052 +  0.061 +
                  0.000 +  0.000 +
                  0.018 +  0.021 = 0.153
      df = 2
```

p-value = 0.9265

For $\alpha = 0.05$, the critical value is $\chi^2_{0.05} = 5.99$. The observed χ^2 is not significant. As such, the null hypothesis of independence between period of acquaintanceship and duration of marriage is not contradicted.

13.41 We test the null hypothesis of independence. The χ^2 statistic is calculated (using Minitab) as follows:

```
Expected counts are printed below observed counts

             One   Neither    Total
 Innoc       27        20        47
             22.43     24.57

 Unbal       36        49        85
             40.57     44.43

Total        63        69       132

  Chi-Sq =  0.930 +   0.849 +
            0.514 +   0.470 = 2.764
DF = 1, P-Value = 0.096
```

We take $\alpha = 0.05$. For d.f. = 1, we find that $\chi^2_{0.05} = 3.84$, so that the rejection region is $R: \chi^2 \geq 3.84$. Since the observed value of $\chi^2 = 2.764$ does not lie in R, we do not reject H_0 at $\alpha = 0.05$.

13.43 (a) In Table, 15 the expected frequency of the first cell is $75\left(\frac{100}{300}\right) = 25$, which is the same as the observed frequency. In the same manner, it is seen that the expected and observed frequencies are identical in every cell. Consequently, $\chi^2 = 0$. This is also the case for Table 16.

(b) The χ^2 statistic is calculated (using Minitab) for the pooled data in Table 17 as follows:

p-value < 0.001.

For d.f. = 1, we find that $\chi^2_{0.05} = 3.84$, so the null hypothesis is rejected at $\alpha = 0.05$.

(c) Note that for secretarial positions the proportion of candidates receiving an offer is 25/75 = 1/3 for males and 75/225 = 1/3 for females. The proportions for sales positions are 150/200 = 3/4 for males and 75/100 = 3/4 for females. Although the rates are equal within each table, we see that in the pooled table, the rates are 175/ 275 = 0.64 for males and 150/325 = 0.46 for females. This discrepancy arises because in Table 15, 75/300 (or 25%) are male applicants, whereas in Table 16 there are 67%. Thus, pooling the tables results in a rather uneven mix. Since the overall rates of offer are very different for the two kinds of positions, the tables should not be combined.

Chapter 14

ANALYSIS OF VARIANCE (ANOVA)

14.1 (a) We first find $k = 4,\ \bar{y} = 6,\ \bar{y}_1 = 7,\ \bar{y}_2 = 6,\ \bar{y}_3 = 3$, and $\bar{y}_4 = 8$. Thus,

$$\begin{matrix} \text{Obs.} & & \text{Grand mean} & & \text{Tr. Effect} & & \text{Residuals} \\ y_{ij} & & \bar{y} & & \bar{y}_i - \bar{y} & & y_{ij} - \bar{y}_i \\ \begin{bmatrix} 5 & 9 \\ 8 & 4 \\ 4 & 2 \\ 7 & 9 \end{bmatrix} & = & \begin{bmatrix} 6 & 6 \\ 6 & 6 \\ 6 & 6 \\ 6 & 6 \end{bmatrix} & + & \begin{bmatrix} 1 & 1 \\ 0 & 0 \\ -3 & -3 \\ 2 & 2 \end{bmatrix} & + & \begin{bmatrix} -2 & 2 \\ 2 & -2 \\ 1 & -1 \\ -1 & 1 \end{bmatrix} \end{matrix}$$

(b) Treatment SS $= 2(1)^2 + 2(0)^2 + 2(-3)^2 + 2(2)^2 = 28$
Residual SS $= (-2)^2 + (2)^2 + \ldots + (-1)^2 + 1^2 = 20$
Mean SS $= 8(6)^2 = 288$
Total SS $= 5^2 + 9^2 + \ldots + 9^2 = 336$
Total SS (corrected) $= (5-6)^2 + \ldots + (9-6)^2 = 336 - 288 = 48$

(c) Error d.f. $= \sum n_i - k = 2 + 2 + 2 + 2 - 4 = 4$
Treatment d.f. $= k - 1 = 4 - 1 = 3$

(d) The analysis-of-variance table is

ANOVA Table

Source	Sum of Squares	d.f.
Treatment	28	3
Error	20	4
Total	48	7

14.3 (a) We first find $k = 3$, $\bar{y}_1 = 5$, $\bar{y}_2 = 3$, $\bar{y}_3 = 1$, and $\bar{y} = 33/11 = 3$. Thus,

Obs.	Grand mean	Tr. Effect	Residuals
y_{ij}	$\bar{y}$	$\bar{y}_i - \bar{y}$	$y_{ij} - \bar{y}_i$

$$\begin{bmatrix} 7 & 5 & 4 & 4 \\ 6 & 1 & 2 & \\ 2 & 1 & 0 & 1 \end{bmatrix} = \begin{bmatrix} 3 & 3 & 3 & 3 \\ 3 & 3 & 3 & \\ 3 & 3 & 3 & 3 \end{bmatrix} + \begin{bmatrix} 2 & 2 & 2 & 2 \\ 0 & 0 & 0 & \\ -2 & -2 & -2 & -2 \end{bmatrix} + \begin{bmatrix} 2 & 0 & -1 & -1 \\ 3 & -2 & -1 & \\ 1 & 0 & -1 & 0 \end{bmatrix}$$

(b) Treatment SS $= 4(2^2) + 3(0^2) + 4(-2)^2 = 32$
Residual SS $= 2^2 + 0^2 + \ldots + 0^2 = 22$
Mean SS $= 11(3)^2 = 99$
Total SS $= 7^2 + 5^2 + \ldots 0^2 + 1^2 = 153$
Total SS (corrected) $= (7-3)^2 + (5-3)^2 + (4-3)^2 + (4-3)^2 + \ldots + (1-3)^2 = 153 - 99 = 54$

(c) Treatment d.f. $= k - 1 = 3 - 1 = 2$
Residual d.f. $= \sum n_i - k = 4 + 3 + 4 - 3 = 8$
Total d.f. $= 4 + 3 + 4 - 1 = 10$

(d) The analysis-of-variance table is

ANOVA Table

Source	Sum of Squares	d.f.
Treatment	32	2
Error	22	8
Total	54	10

14.5 We first find $\bar{y}_1 = 2$, $\bar{y}_2 = 3$, $\bar{y}_3 = 6$, $\bar{y}_4 = 4$, and $\bar{y} = \frac{48}{12} = 4$. Thus,

Obs.	Grand mean	Tr. Effect	Residuals
y_{ij}	$\bar{y}$	$\bar{y}_i - \bar{y}$	$y_{ij} - \bar{y}_i$

$$\begin{bmatrix} 2 & 1 & 3 & \\ 1 & 5 & & \\ 9 & 5 & 6 & 4 \\ 3 & 4 & 5 & \end{bmatrix} = \begin{bmatrix} 4 & 4 & 4 & \\ 4 & 4 & & \\ 4 & 4 & 4 & 4 \\ 4 & 4 & 4 & \end{bmatrix} + \begin{bmatrix} -2 & -2 & -2 & \\ -1 & -1 & -1 & \\ -2 & 2 & 2 & 2 \\ 0 & 0 & 0 & \end{bmatrix} + \begin{bmatrix} 0 & -1 & 1 & \\ -2 & 2 & & \\ 3 & -1 & 0 & -2 \\ -1 & 0 & 1 & \end{bmatrix}$$

Treatment SS $= 3(-2)^2 + 2(-1)^2 + 4(-2^2) + 3(0)^2 = 30$
Residual SS $= 0^2 + (-1)^2 + 1^2 + \ldots + 0^2 + 1^2 = 26$
Total SS $= (2-4)^2 + (1-4)^2 + (3-4)^2 + (1-4)^2 + \ldots + (5-4)^2 = 56$
Treatment d.f. $= k - 1 = 4 - 1 = 3$
Residual d.f. $= \sum n_i - k = 3 + 2 + 4 + 3 - 4 = 8$
Total d.f. $= 3 + 2 + 4 + 3 - 1 = 11$

The analysis-of-variance table is

ANOVA Table

Source	Sum of Squares	d.f.
Treatment	30	3
Error	26	8
Total	56	11

14.7 (a) $SSE = (n_1 - 1)S_1^2 + (n_2 - 1)S_2^2 + (n_3 - 1)S_3^2 = 3018.81$.

(b) We have the following ANOVA table:

Source	Sum of Squares	df	Mean Square	F-ratio
Treatment	346.792	2	173.36	
Error	3018.81	98	30.804	
Total	3365.593	100		5.628

Note that

$$\text{mean square} = \text{SS/df} = 173.36,$$

and

$$F = \frac{\frac{\text{Treatment SS}}{2}}{\frac{\text{Error SS}}{\sum n_i - 3}} = 5.628.$$

(c) Using $\alpha = 0.05$, the rejection region is R: $F \geq F(2,98) = \frac{3.07+4.98}{2} = 4.025$. Since our test statistic of 5.628 is in the rejection region, we reject the null hypothesis that the means are equal. So, the means differ across groups.

14.9 (a) $F_{0.10}(3, 5) = 3.62$
(b) $F_{0.10}(3, 10) = 2.73$
(c) $F_{0.10}(3, 15) = 2.49$
(d) $F_{0.10}(3, 30) = 2.28$
(e) Increasing the denominator d.f. decreases the upper 10th percentile.

14.11 For $v_1 = 5$ and $v_2 = 30$ in the F-table, $F_{0.05}(5, 30) = 2.53$. We observe that

$$F = \frac{\text{Treatment } SS/(k-1)}{SSE/(n-k)} = \frac{23/5}{56/30} = 2.46$$

So, we fail to reject $H_0: \mu_1 = \mu_2 = \mu_3 = \mu_4 = \mu_5 = \mu_6$, at level $\alpha = 0.05$.

14.13 We are to test the null hypothesis $H_0: \mu_1 = \mu_2 = \mu_3$ versus the alternative hypothesis that the means are not all equal. Given $\alpha = 0.05$, the rejection region is determined by the value $F_{0.05}(2, 9) = 4.26$ obtained from the F-table. From Exercise 14.2, the observed value of F is

$$F = \frac{\text{Treatment } SS/(k-1)}{SSE/(n-k)} = \frac{312/2}{170/9} = 8.26$$

Consequently, we reject the null hypothesis that the means are equal at the $\alpha = 0.05$ level of significance.

14.15 We are to test the null hypothesis $H_0 : \mu_1 = \mu_2 = \mu_3$ versus the alternative hypothesis that the means are not all equal. Given $\alpha = 0.05$, the rejection region is determined by the value $F_{0.05}(2, 98) = 2.36$ obtained from the F-table. From Exercise 14.6, the observed value of F is

$$F = \frac{\text{Treatment } SS/(k-1)}{SSE/(n-k)} = 5.628$$

Consequently, we reject H_0 at the $\alpha = 0.05$ level.

14.17 For multiple-t confidence intervals we use $t_{\alpha/2m}$ with d.f. $= n-k$.

(a) $\frac{\alpha}{2m} = \frac{0.05}{2(3)} = 0.00833$ and, with d.f. = 26, $t_{0.0083} = 2.559$

(b) $\frac{\alpha}{2m} = \frac{0.05}{2(5)} = 0.005$ and, with df = 26, $t_{0.005} = 2.779$

14.19 The error d.f. = (14 + 20 + 30 +15 - 4) = 75. The t intervals use $t_{0.5} = 1.645$ and the multiple-t intervals use $\frac{\alpha}{2m} = \frac{0.10}{2(6)} = 0.0083$ so we extrapolate $t_{0.0083} = 2.576$

Using the formulae $(\bar{y}_i - \bar{y}_j) \pm 1.645(1.56)\sqrt{\frac{1}{n_i} + \frac{1}{n_j}}$ for the t-interval and

$(\bar{y}_i - \bar{y}_j) \pm 2.576(1.56)\sqrt{\frac{1}{n_i} + \frac{1}{n_j}}$ for the multiple t-interval, we obtain:

	(a) t-interval	(b) multiple-t interval
$\mu_1 - \mu_2$:	0.9 ± 2.614	0.9 ± 4.094
$\mu_1 - \mu_3$:	0.37 ± 0.831	0.37 ± 1.301
$\mu_1 - \mu_4$:	-0.06 ± 0.954	-0.06 ± 1.494
$\mu_2 - \mu_3$:	-0.53 ± 0.704	-0.53 ± 0.116
$\mu_2 - \mu_4$:	-0.96 ± 0.843	-0.96 ± 1.320
$\mu_3 - \mu_4$:	-0.43 ± 0.812	-0.43 ± 1.272

None of the population means differ according to the multiple t-intervals.

14.21 The t-interval is given by $\overline{Y}_i - \overline{Y}_{i'} \pm t_{\alpha/2} S\sqrt{\frac{1}{n_i} + \frac{1}{n_{i'}}}$, while the multiple-$t$ interval is given by $\overline{Y}_i - \overline{Y}_{i'} \pm t_{\alpha/2m} S\sqrt{\frac{1}{n_i} + \frac{1}{n_{i'}}}$. So, the ratio of their lengths, namely,

$$\frac{2t_{\alpha/2} S\sqrt{\frac{1}{n_i} + \frac{1}{n_{i'}}}}{2t_{\alpha/2m} S\sqrt{\frac{1}{n_i} + \frac{1}{n_{i'}}}} = \frac{t_{\alpha/2}}{t_{\alpha/2m}}$$

does not depend on the data. For $m = 10$ and $\alpha = 0.10$, this ratio is

$$\frac{t_{0.05}}{t_{0.005}} = \frac{1.753}{2.947} = 0.595 \text{ for d.f.} = 15.$$

14.23 (a)

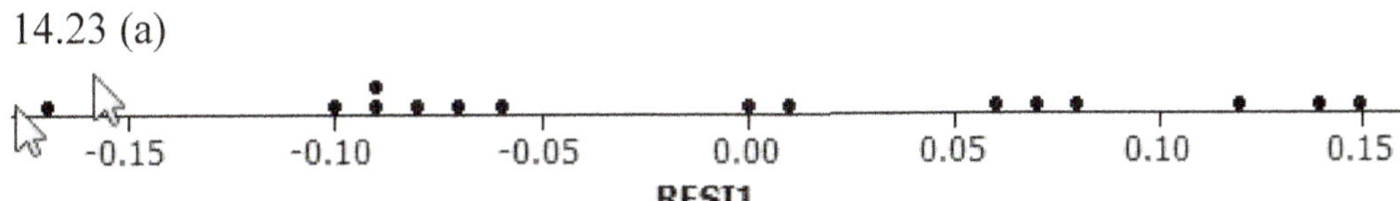

14.23(b) Recipe 1 on top.

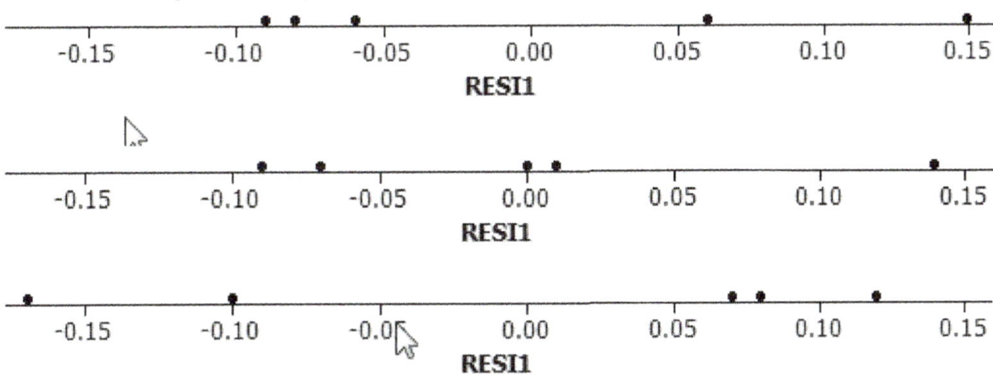

14.25 (a) We first find $k = 3$, $\overline{y}_{..} = 8$, $\overline{y}_{1.} = 7$, $\overline{y}_{2.} = 6$, $\overline{y}_{3.} = 11$. Also, $b = 4$, $\overline{y}_{.1} = 11$, $\overline{y}_{.2} = 8$, $\overline{y}_{.3} = 9$, and $\overline{y}_{.4} = 4$. Thus,

$$\underset{\substack{\text{Obs.}\\ y_{ij}}}{\begin{bmatrix} 11 & 10 & 7 & 0 \\ 7 & 8 & 7 & 2 \\ 15 & 6 & 13 & 10 \end{bmatrix}} = \underset{\substack{\text{Grand mean}\\ \overline{y}_{..}}}{\begin{bmatrix} 8 & 8 & 8 & 8 \\ 8 & 8 & 8 & 8 \\ 8 & 8 & 8 & 8 \end{bmatrix}} + \underset{\substack{\text{Tr. Effect}\\ \overline{y}_{i.} - \overline{y}_{..}}}{\begin{bmatrix} -1 & -1 & -1 & -1 \\ -2 & -2 & -2 & -2 \\ 3 & 3 & 3 & 3 \end{bmatrix}}$$

$$+ \underset{\substack{\text{Bl. Effect}\\ \overline{y}_{.j} - \overline{y}_{..}}}{\begin{bmatrix} 3 & 0 & 1 & -4 \\ 3 & 0 & 1 & -4 \\ 3 & 0 & 1 & -4 \end{bmatrix}} + \underset{\substack{\text{Error}\\ y_{ij} - \overline{y}_{i.} - \overline{y}_{.j} + \overline{y}_{..}}}{\begin{bmatrix} 1 & 3 & -1 & -3 \\ -2 & 2 & 0 & 0 \\ 1 & -5 & 1 & 3 \end{bmatrix}}$$

(b) The sums of squares are:

Treatment SS $= 4\,(-1)^2 + 4(-2)^2 + 4(3)^2 = 56$
Block SS $= 3(3)^2 + 3(0)^2 + 3(1)^2 + 3(-4)^2 = 78$
Residual SS $= 1^2 + 3^2 + (-1)^2 + \ldots + 1^2 + 3^2 = 64$
Mean $= 12(8)^2 = 768$
Total SS $= 11^2 + 10^2 + 7^2 + \ldots + 13^2 + 10^2 = 966$
Total SS (corrected $= 11^2 + 10^2 + 7^2 + \ldots + 13^2 + 10^2 - 12\,(8)^2$
$= 966 - 768 = 198$

(c) Treatment d.f. $= k - 1 = 2$
Block d.f. $= b - 1 = 3$
Residual d.f. $= (k - 1)\,(b - 1) = (2)(3) = 6$

14.27 The null hypothesis is that the three treatment population means are the same; the alternative hypothesis is that they are not the same. The analysis-of-variance table is

Source of variation	Degrees of freedom	Sum of squares	Mean square	F
Treatments	2	56	28	2.62
Blocks	3	78	26	2.44
Error	6	64	10.667	
Total	11	198		

Since the critical value at the 0.05 level for an F distribution with 2 and 6 degrees of freedom is 5.14, we fail to reject the null hypothesis of equal treatment means. Since $F_{0.05}$ with 3 and 6 degrees of freedom is 4.76, the block effect is not significant.

14.29 (a) At each baking, select a loaf of bread from each recipe and randomize the position of the loaves in the oven.

(b) The Grand mean is $\bar{y}_{..} = \frac{1}{15}(0.95 + 0.71 + ... + 0.44) = \frac{10.77}{15} = 0.718$.
The Block means are 0.7833, 0.863, 0.6133, 0.7233, and 0.6067.
The recipe means are 0.796, 0.708, and 0.650.

$SS_B = 3[(0.7833 - 0.718)^2 + (0.863 - 0.718)^2 + (0.6133 - 0.718)^2 + (0.7233 - 0.718)^2 + (0.6067 - 0.718)^2] = 0.1015$
$SS_T = 5[(0.796 - 0.718)^2 + (0.708 - 0.718)^2 + (0.650 - 0.718)^2] = 0.0540$
Total $SS = (0.95 - 0.718)^2 + \ldots + (0.44 - 0.718)^2 = 0.2971$
Hence, we have $SSE = 0.2971 - (0.0540 + 0.1015) = 0.1416$
The ANOVA table is

Source of variation	Sum of Squares	df	Mean square	F
Treatments	0.0540	2	0.027	8.17
Blocks	0.1015	4	0.0254	4.66
Residual	0.1416	8	0.0177	
Total	0.2971	14		

Since $F_{0.05}(2, 8) = 4.46 < 8.17$ we conclude that a significant treatment difference is indicated by the data. Also, $F_{0.05}(4, 8) = 3.84 < 4.66$, so the block effects are significant.

14.31 (a) The Grand mean is
$$\bar{y}_{..} = \frac{1}{36}(19.09 + 16.28 + ... + 21.58) = \frac{653.96}{36} = 18.1656.$$
The block means are 17.7533, 18.78, 18.2367, 17.94, 17.6867, and 18.5967.
The variety means are 19.6683, 17.1083, 17.2683, 17.7, 16.0767, and 21.177.
$SS_B = 6[(17.7533 - 18.1656)^2 + (18.78 - 18.1656)^2 + \ldots + (18.5967 - 18.1656)^2] = 6.112$
$SS_T = 6[(19.6683 - 18.1656)^2 + (17.1083 - 18.1656)^2 + \ldots + (21.1717 - 18.1656)^2] = 106.788$
$SS = (19.09 - 18.1656)^2 + (16.28 - 18.1656)^2 + \ldots + (21.58 - 18.1656)^2 = 117.898$

Hence, we have $SSE = 117.898 - (6.112 + 106.788) = 4.998$.
The ANOVA table is

Source of variation	Sum of squares	d.f.	Mean square	F-ratio
Treatments	106.788	5	21.358	106.79
Blocks	6.112	5	1.222	6.11
Error	4.998	25	0.200	
Total	117.898	35		

Since $F_{0.05}(5, 25) = 2.603 < 106.79$ we conclude that a highly significant treatment difference is indicated by the data. Also, $F_{0.05}(5, 25) < 6.11$ so the block effects are significant.

(b) Array of residuals

$$\begin{bmatrix} -0.166 & -0.416 & -0.546 & 0.212 & 0.586 & 0.331 \\ 0.007 & 0.157 & 0.287 & -0.264 & 0.229 & -0.416 \\ 0.571 & -0.299 & 0.041 & -0.181 & -0.268 & 0.137 \\ 0.157 & 0.687 & 0.487 & 0.166 & -1.071 & -0.426 \\ -0.569 & 0.091 & -0.449 & 0.159 & 0.372 & 0.397 \\ 0.001 & -0.219 & 0.181 & -0.091 & 0.151 & -0.023 \end{bmatrix}$$

The dot diagrams are given on the next page. Note that -1.071 is a possible outlier. The normal scores plot is also given on the next page (directly following the dot diagrams) and the same point is a little low.

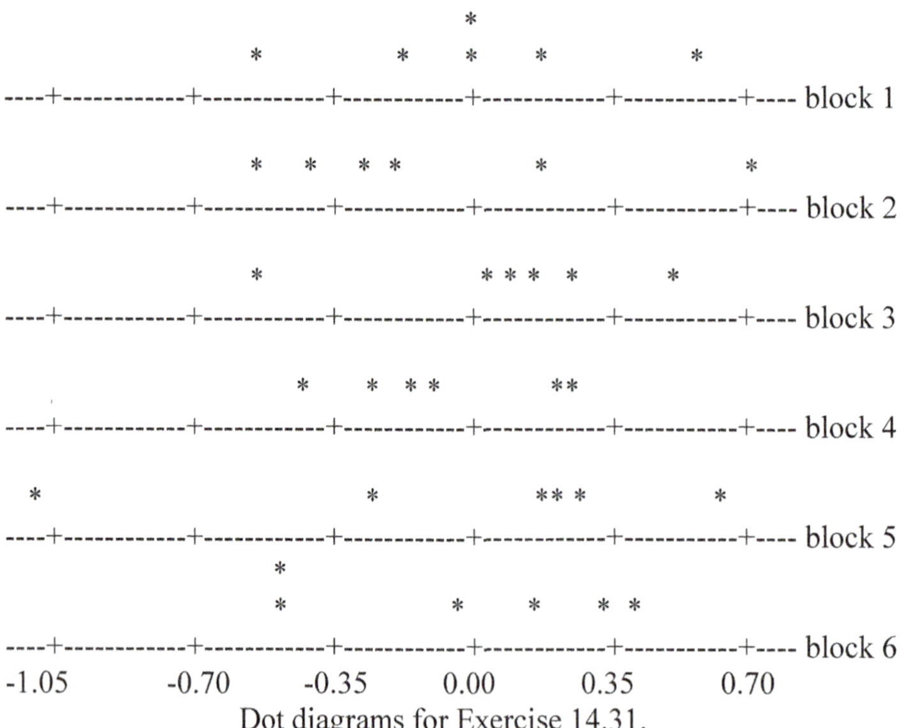

Dot diagrams for Exercise 14.31.

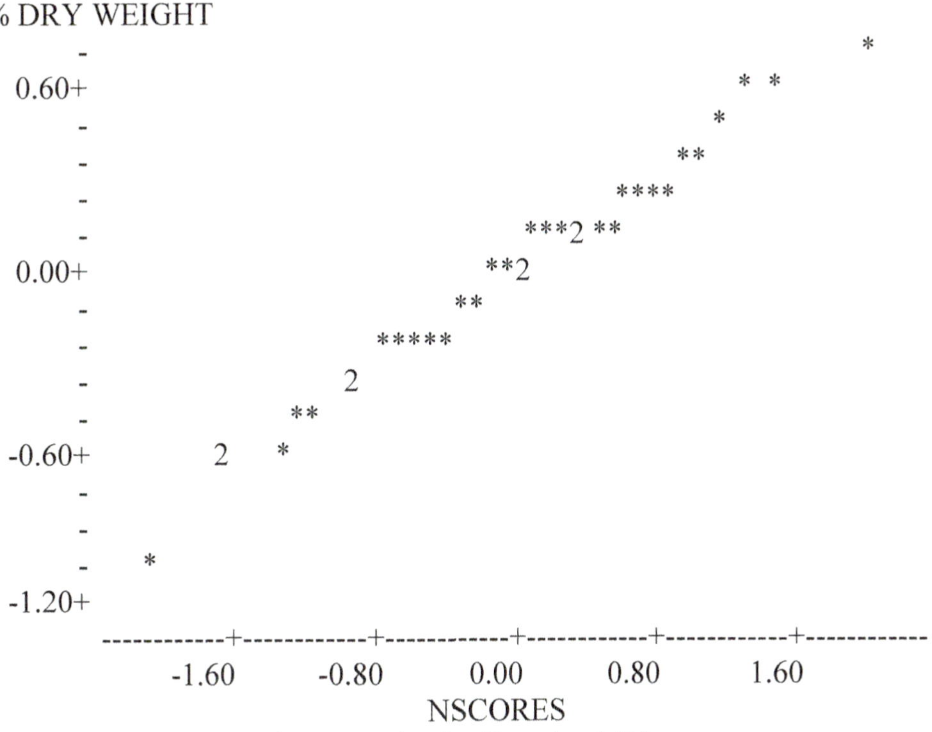

Normal-scores plot for Exercise 1431.

14.33 Note that $k = 3$. We first find $\overline{y}_1 = 19, \overline{y}_2 = 13, \overline{y}_3 = 14, \overline{y} = 225/15 = 15$.

$$\underset{y_{ij}}{\begin{bmatrix} 19 & 18 & 21 & 18 & & \\ 16 & 11 & 13 & 14 & 11 & \\ 13 & 16 & 18 & 11 & 15 & 11 \end{bmatrix}} = \underset{\overline{y}}{\begin{bmatrix} 15 & 15 & 15 & 15 & & \\ 15 & 15 & 15 & 15 & 15 & \\ 15 & 15 & 15 & 15 & 15 & 15 \end{bmatrix}}$$

$$+ \underset{\overline{y}_i - \overline{y}}{\begin{bmatrix} 4 & 4 & 4 & 4 & & \\ -2 & -2 & -2 & -2 & -2 & \\ -1 & -1 & -1 & -1 & -1 & -1 \end{bmatrix}} + \underset{y_{ij} - \overline{y}_i}{\begin{bmatrix} 0 & -1 & 2 & -1 & & \\ 3 & -2 & 0 & 1 & -2 & \\ -1 & 2 & 4 & -3 & 1 & -3 \end{bmatrix}}$$

14.35 We chose $\alpha = 0.05$. With $m = \binom{3}{2} = 3$, we have $\frac{\alpha}{2m} = 0.0083$. From Appendix Table 4 with d.f. = 2, we interpolate $t_{0.0083} = 6.965$. From the ANOVA table, $s = \sqrt{30.804} \approx 5.55$. Hence simultaneous 95% confidence intervals for the differences are given by

$\mu_1 - \mu_2:\quad (5.27 - 9.62) \pm 6.965 \times 5.55\sqrt{\frac{1}{35} + \frac{1}{33}} = -4.35 \pm 9.016$

or $(-13.366, 4.716)$

$\mu_1 - \mu_3:\quad (5.27 - 8.45) \pm 6.965 \times 5.55\sqrt{\frac{1}{35} + \frac{1}{33}} = -3.18 \pm 9.016$

or $(-12.196, 5.836)$

$\mu_2 - \mu_3$: $(9.62 - 8.45) \pm 6.965 \times 5.55\sqrt{\frac{1}{33} + \frac{1}{33}} = 1.17 \pm 9.147$
or $(-7.977, 10.317)$

(b) None of the means is statistically different from either of the other two. This is consistent with the observation in the earlier exercise.

14.37 The Minitab output is shown below:

```
One-way Analysis of Variance

Analysis of Variance
Source    DF       SS       MS       F        P
Factor     2    5.279    2.640   20.85    0.000
Error     14    1.772    0.127
Total     16    7.052
    Individual 95% CIs For Mean

Level      N    Mean      StDev   --+---------+---------+---------+-
Treat 1    5  1.4600     0.1306   (----*-----)
Treat 2    6  1.5967     0.4357     (-----*----)
Treat 3    6  2.6950     0.3886                       (----*----)
                                  --+---------+---------+---------+-
Pooled StDev =   0.3558          1.20      1.80      2.40      3.00
                                            Moisture
```

14.39 (a) We first find $k = 3$, $\bar{y}_{..} = 8$, $\bar{y}_{1.} = 6$, $\bar{y}_{2.} = 7$, $\bar{y}_{3.} = 11$. Also, $b = 4$, $\bar{y}_{.1} = 7$, $\bar{y}_{.2} = 12$, $\bar{y}_{.3} = 3$, and $\bar{y}_{.4} = 10$. Thus,

$$\underset{y_{ij}}{\overset{\text{Obs.}}{\begin{bmatrix} 8 & 9 & 1 & 6 \\ 5 & 12 & 0 & 11 \\ 8 & 15 & 8 & 13 \end{bmatrix}}} = \underset{\bar{y}_{..}}{\overset{\text{Grand mean}}{\begin{bmatrix} 8 & 8 & 8 & 8 \\ 8 & 8 & 8 & 8 \\ 8 & 8 & 8 & 8 \end{bmatrix}}} + \underset{\bar{y}_{i.} - \bar{y}_{..}}{\overset{\text{Tr. Effect}}{\begin{bmatrix} -2 & -2 & -2 & -2 \\ -1 & -1 & -1 & -1 \\ 3 & 3 & 3 & 3 \end{bmatrix}}}$$

$$+ \underset{\bar{y}_{.j} - \bar{y}_{..}}{\overset{\text{Bl. Effect}}{\begin{bmatrix} -1 & 4 & -5 & 2 \\ -1 & 4 & -5 & 2 \\ -1 & 4 & -5 & 2 \end{bmatrix}}} + \underset{y_{ij} - \bar{y}_{i.} - \bar{y}_{.j} + \bar{y}_{..}}{\overset{\text{Error}}{\begin{bmatrix} 3 & -1 & 0 & -2 \\ -1 & 1 & -2 & 2 \\ -2 & 0 & 2 & 0 \end{bmatrix}}}$$

(b) The sums of squares are:

Treatment SS	$= 4\,(-2)^2 + 4(-1)^2 + 4(3)^2 = 56$
Block SS	$= 3(-1)^2 + 3(4)^2 + 3(-5)^2 + 3(2)^2 = 138$
Residual SS	$= 3^2 + (-1)^2 + \ldots + 2^2 + 0^2 = 32$
Total SS	$= 8^2 + 9^2 + 1^2 + \ldots + 8^2 + 13^2 - 12(8)^2 = 226$

(c) The $k = 3$ rows of the treatment array sum to zero, and the $b = 4$ columns of the block array sum to zero. All of the entries of the Error array sum to zero as do the columns and the rows. Consequently,

Treatment d.f. $= k - 1 = 2$
Block d.f. $= b - 1 = 3$
Residual d.f. $= (k - 1)\,(b - 1) = (2)(3) = 6$

Chapter 15

NONPARAMETRIC INFERENCE

15.1 (a) Rank collections for treatment B with sample sizes $n_A = 4$ and $n_B = 2$

Rank of B	Rank Sum W_B	Probability
1,2	3	1/15
1,3	4	1/15
1,4	5	1/15
1,5	6	1/15
1,6	7	1/15
2,3	5	1/15
2,4	6	1/15
2,5	7	1/15
2,6	8	1/15
3,4	7	1/15
3,5	8	1/15
3,6	9	1/15
4,5	9	1/15
4,6	10	1/15
5,6	11	1/15
	Total	1

Under H_0, when the two samples come from the same population, every pair of integers out of {1,2,3,4,5,6} is equally likely to be the ranks for the two B measurements. There are $\binom{6}{2} = 15$ potential pairs, so that each collection of possible ranks has a probability of 1/15.

(b) Both rank collections {1,4} and {2,3} have $W_B = 5$ so

$$P[W_B = 5] = 1/15 + 1/15 = 2/15.$$

Continuing, we obtain the distribution of W_B:

Values of W_B	3	4	5	6	7	8	9	10	11
Probability	$\frac{1}{15}$	$\frac{1}{15}$	$\frac{2}{15}$	$\frac{2}{15}$	$\frac{3}{15}$	$\frac{2}{15}$	$\frac{2}{15}$	$\frac{1}{15}$	$\frac{1}{15}$

These values agree with the tabulated entries of Table 7.

15.3 (a) Smaller sample size = 5, larger sample size = 6, so $P[W_S \geq 39] = 0.063$

(b) Smaller sample size = 4, larger sample size = 6, so $P[W_S \leq 15] = 0.086$

(c) Smaller sample size = 7, larger sample size = 7, $P[W_S \geq 66] = 0.049$, so c = 66.

15.5

(a) Since n_A is the smaller sample size, $W_A = 2 + 4 + 5 = 11$.

(b)

Combined sample ordered observations	0.41	2.42	2.58	7.61	10.39
Ranks	1	2	3	4	5
Treatment	A	B	A	B	B

We find $W_S = 1 + 3 = 4$.

15.7 We test H_0 : the populations are identical versus H_1 : the populations are different. Let W_S = rank sum of phosphate for the Chester White breed. The alternative is two-sided. From Appendix Table 9 we find with smaller size = 8 = larger size, $P[W_S \geq 87] = 0.025 = P[W_S \leq 49]$.

The combined ordered observations, with the Chester White underlined, are

Ordered observations	<u>47</u>	<u>48</u>	<u>57</u>	58	<u>65</u>	<u>75</u>	78	79
Ranks	1	2	<u>3</u>	4	<u>5</u>	<u>6</u>	7	8

Ordered observations	<u>97</u>	<u>99</u>	<u>110</u>	162	172	182	220	230
Ranks	<u>9</u>	<u>10</u>	<u>11</u>	12	13	14	15	16

We find $W_S = 1 + 2 + 3 + 5 + 6 + 9 + 10 + 11 = 47$

We conclude that the serum phosphate level is significantly different for the two breeds at the level $\alpha = 0.05$.

15.9 We test H_0: the populations are identical versus H_1: the populations are different. Let W_S = rank sum for treatment 2. The alternative is two-sided. From Appendix Table 9 we find with smaller size = 7 and larger size = 8, $P[WS \geq 73] = 0.027 = P[WS \leq 39]$. The sample sizes $n_A = 8$ and $n_B = 7$.

Combined ordered values	18	<u>25</u>	28	29	<u>30</u>	31	<u>36</u>	<u>37</u>	38	40	<u>41</u>	43	46	<u>49</u>	<u>56</u>
Ranks	1	<u>2</u>	3	4	<u>5</u>	6	<u>7</u>	<u>8</u>	9	10	<u>11</u>	12	13	<u>14</u>	<u>15</u>

$$W_S = 2 + 5 + 7 + 8 + 11 + 14 + 15 = 62$$

Consequently, we fail to reject the null hypothesis at level $\alpha = 0.054$.

15.11 (a) The configuration that most supports the alternative hypothesis is *BBBBBBBBBA* where the rank of the single *A* observation is 10.
(b) There are 10 possible ranks (positions) for the single *A* and these are equally likely. Therefore, $P[W_A = 10] = 0.1$.
(c) The single most extreme outcome has probability 0.1. An α of 0.05 cannot be achieved unless, whenever $W_A = 10$, we are willing to flip a coin to decide whether or not H_0 should be rejected.

15.13 The alternative is one-sided. We test H_0: there is no difference in treatments against H_1: $P[+] > 0.5$. Let S = number of positive signs among the differences. The rejection region is $R : S \geq c$. There are 13 positive values out of 18. From Appendix Table 2 with $n = 18$ we find $P[S \geq 13] = 0.048$. Since the observed value is $S = 13$, we reject H_0, at level $\alpha = 0.048$, in favor of $P[+] > 0.5$ or more than half of the population prefer recipe *A*.

15.15 The alternative is two-sided. We test H_0: no difference in treatments against H_1: $P[+] = 0.5$. Let S = number of positive signs among the differences. The rejection region is $R : S \leq c_1$ or $S \geq c_2$. Since there are 7 ties, the effective sample size is $25 - 7 = 18$. From Appendix Table 2 with $n = 18$, we find $P[S \leq 4] = 0.015 = P[S \geq 14]$. Since the observed value is $S = 11$, we conclude at $\alpha = 0.030$ that the data do not show a significant difference of opinion between husbands and wives.

15.17 (a) $P[T^+ \geq 65] = 0.021$,
(b) $P[T^+ \leq 10] = 0.042$
(c) Since $P[T^+ \geq 31] = 0.039$, we have $c = 31$.
(d) Since $P[T^+ \geq 55] = 0.027 = P[T^+ \leq 11]$, we have $c_1 = 11$ and $c_2 = 55$.

15.19 (a)

	Ranks 1, 2, 3	$T+$	Probability
	+, +, +	6	0.125
	+, +, -	3	0.125
	+, -, +	4	0.125
Signs	-, +, +	5	0.125
	+, -, -	1	0.125
	-, +, -	2	0.125
	-, -, +	3	0.125
	-, -, -	0	0.125

(b)

Values of T^+	0	1	2	3	4	5	6
Probability	0.125	0.125	0.125	0.250	0.125	0.125	0.125

From the table when $n = 3$, we see that the tail probabilities agree.

15.21 (a) The alternative is one-sided. We test H_0: there is no difference in treatments against H_1: $P[+] > 0.5$. There are two ties among the differences so the effective sample size is $n = 13$. Let S = number of positive signs among the 13 differences *before – after*. The rejection region is $R : S > c$. From Appendix Table 2 with $n = 13$ we find $P[S \geq 10] = 0.046$.

(b) The alternative is one-sided so we reject for large values of T_+. Using the 0 differences for ranking and average rank for the other ties, we calculate

Ordered absolute value of differences	0	0	2	4	4	6	8	8
Ranks	1.5	1.5	3	4.5	4.5	6	7.5	7.5
Signs			+	+	-	+	-	+

Ordered absolute value of differences	10	10	10	18	18	26	32
Ranks	10	10	10	12.5	12.5	14	15
Signs	+	+	+	+	+	+	+

The resulting observed value of T^+ is 105. If there were no ties, $R : T^+ \geq 990$ since $P[T^+ \geq 90] = 0.047$ for $n = 15$. Consequently, we reject H_0 and conclude that the mean blood pressure has been reduced.

15.23 We calculate

x	64.9	77.0	51.4	51.0
y	0.813	0.806	0.783	0.771
Ranks R_i	3	4	1	2
Ranks X_i	3	4	2	1

$$r_{S_p} = \frac{\sum_{i=1}^{4}\left(R_i - \frac{4+1}{2}\right)\left(X_i - \frac{4+1}{2}\right)}{\frac{4\left(4^2-1\right)}{12}}$$

$$= \frac{(1-2.5)(2-2.5)+(2-2.5)(1-2.5)+(3-2.5)(3-2.5)+(4-2.5)(4-2.5)}{5} = 0.7$$

15.25 We are to test H_0 : independence against a two-sided alternative.

Student	1	2	3	4	5	6	7	8	9	10
Dexterity	23	29	45	36	49	41	30	15	42	38
Aggression	45	48	16	28	38	21	36	18	31	37
Ranks R_i	2	3	9	5	10	7	4	1	8	6
Ranks S_i	9	10	1	4	8	3	6	2	5	7

$$r_{S_p} = \frac{\sum_{i=1}^{10}\left(R_i - \frac{10+1}{2}\right)\left(S_i - \frac{10+1}{2}\right)}{\frac{10\left(10^2-1\right)}{12}}$$

$$= \frac{(2-5.5)(9-5.5)+(3-5.5)(10-5.5)+\cdots+(6-5.5)(7-5.5)}{82.5}$$

$$= -\frac{16.5}{82.5} = -0.20$$

Even though $n = 10$ is not large, we approximate that $\sqrt{n-1}\, r_{Sp}$ is approximately standard normal. Since the z value $\sqrt{9}\,(-0.20) = -0.60$ is not negative enough, we cannot reject the null hypothesis of independence.

15.27

Combined sample ordered observations	8	4	6	1	5	3
Ranks	6	3	5	1	4	2
Treatment	A	A	A	B	B	B

We find $W_A = 6+3+5=14$

15.29 (a) Rank collections for treatment A with sample sizes $n_A = 3$ and $n_B = 2$

Rank of A	Rank Sum W_A		Probability
1,2,3	6		1/10
1,2,4	7		1/10
1,2,5	8		1/10
1,3,4	8		1/10
1,3,5	9		1/10
1,4,5	10		1/10
2,3,4	9		1/10
2,3,5	10		1/10
2,4,5	11		1/10
3,4,5	12		1/10
		Total	1

When the two samples come from the same population, every triple is equally likely to be the ranks for the three A measurements. There are $\binom{5}{3} = 10$ potential triples, so that each collection of possible ranks has a probability of 1/10.

(b) Both rank collections {1,2,5} and {1,3,4} have $W_A = 8$, so

$$P[W_A = 8] = 1/10 + 1/10 = 0.2.$$

Continuing, we obtain the distribution of W_A:

Values of W_A	6	7	8	9	10	11	12
Probability	0.1	0.1	0.2	0.2	0.2	0.1	0.1

15.31 (a) $P[T^+ \geq 28] = 0.098$,
(b) $P[T^+ \leq 5] = 0.020$
(c) Since $P[T^+ \leq 21] = 0.047$, we have $c = 21$.

15.33 We test H_0: populations A and B are identical versus H_1: they are different.

Combined ordered values	95	98	100	103	104	105	116	127	131	137
Ranks	1	2	3	4	5	6	7	8	9	10

Combined ordered values	140	149	150	151	155	164	167	178	179
Ranks	11	12	13	14	15	16	17	18	19

The rank sum of method 2 (the smaller sample) is

$W_S = 1 + 2 + 3 + 4 + 6 + 7 + 8 + 9 + 11 = 51$.

Referring to the table with smaller sample size = 9 and larger sample size = 10, we find that $P[Ws \geq 69] = 0.047 = P[Ws \geq 111]$. Since the observed value = 51 < 69, we conclude there is a significant difference at level $\alpha = 0.047 + 0.047 = 0.094$. In fact, the null hypothesis would be rejected even for α much smaller than 0.018.

15.35 (a) For the Chester Whites,

Calcium	116	112	82	63	117	69	79	87
Phosphate	47	48	57	75	65	99	97	110
Ranks *Ri*	7	6	4	1	8	2	3	5
Ranks *Si*	1	2	3	5	4	7	6	8

$$r_{S_P} = \frac{\sum_{i=1}^{8}(R_i - 4.5)(S_i - 4.5)}{\frac{8(8^2 - 1)}{12}} = \frac{-22}{42} = -0.524$$

(b) We test the null hypothesis H_0 : independence against a two-sided alternative. The value of the test statistic is $\sqrt{n-1}(-0.524) = -1.386$.

(c) If we approximate that $\sqrt{n-1}r_{S_P}$ is nearly standard normal, the *p*-value would be

$$\alpha = P[Z < -1.386] + P[Z > 1.386] = 0.0829 + 0.0829 = 0.1658$$

This is not less than 0.05, so we would fail to reject the null hypothesis. However, $n = 7$ may not be large enough for a good approximation.

15.37 (a) We calculate

x	166.6	113.8	218.7	149.2	104.9
y	165.7	114.1	217.2	147.3	103.5
difference $d = y - x$	-.9	.3	-1.5	-1.9	-1.4
sign	-	+	-	-	-
Ranks absolute d	2	1	4	5	3

$S = 1$

(b) $T^+ = 1$

CPSIA information can be obtained at www.ICGtesting.com
Printed in the USA
BVOW08s1439120115

382533BV00012B/35/P

9 781118 616314